Eastern Arctic Kayaks

Kayaks and umiaks, East Greenland, c. 1900. Photographer Alfred Berthelsen, no. AI4056.
Courtesy Danish Arctic Institute, Copenhagen.

Eastern Arctic Kayaks

HISTORY, DESIGN, TECHNIQUE

John D. Heath and E. Arima

with contributions by
John Brand, Hugh Collings, Harvey Golden, H. C. Petersen, Johannes Rosing, and Greg Stamer

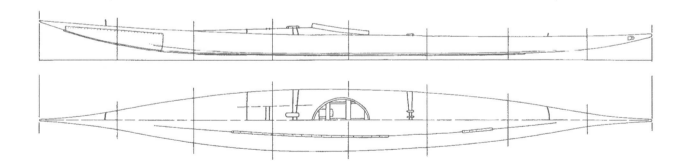

UNIVERSITY OF ALASKA PRESS, FAIRBANKS

University of Alaska Press
P.O. Box 756240
Fairbanks, AK 99775

Library of Congress Cataloging-in-Publication Data:

Heath, John D., 1923–2003.
Eastern Arctic kayaks : history, design, technique / by John D. Heath
and E. Arima with contributions by John Brand ... [et al.].
        p. cm.
    Includes index.
    ISBN 1-889963-25-9 (hardcover : alk. paper)
    1. Kayaks—Greenland. 2. Kayaks—Arctic Coast (Canada).
    3. Kayaks—Canada, Eastern. 4. Inuit—Greenland—Boats.
    5. Inuit—Canada, Eastern—Boats. I. Arima, Eugene Y. II. Title.
VM353.H43 2004
623.82'9—dc22   2004010903

This paper meets the requirements of ANSI/NISO Z39.48–1992
(Permanence of Paper).

Cover and interior design: Dixon Jones, Rasmuson Library Graphics

Printed in the United States
SECOND PRINTING 2006
THIRD PRINTING 2016

Cover photo: Jakob Frederiksen of Nuussuaq, Upernivik region,
Greenland, 1968. Courtesy Keld Hansen.

In Memoriam
John D. Heath
15 April 1923 – 14 July 2003

John Heath was born in Cameron, a small town in central Texas. As a young man during World War II, he volunteered for the Army Air Force. After the war, he became first a draftsman and then a mechanical engineer.

While living in the Seattle area in the early 1950s, John became interested in kayaking. However, he felt that the commercially produced kayaks available to him did not match the design logic of the Native kayaks he saw in museums. Attendance at a lecture featuring Greenland rolls convinced him that recreational paddlers had a lot to learn from the Inuit about paddling skills.

Living in Seattle brought John into contact with Alaska Natives from whom he was able to collect information on techniques, some on the verge of extinction. Upon retirement, he was able to visit Greenland where kayaking was undergoing a revival under the guidance of the legendary Manasse Mathaeussen. John formed a strong relationship with Greenland kayakers, disseminating the information he collected to the wider paddling world via the printed word, videos, and presentations at kayaking symposia in Alaska and elsewhere in the United States, as well as the Channel Islands, France, and Scotland. Perhaps the peak of this involvement was his hosting of the young Greenland Kayak Champion Maligiaq Padilla on a North American tour.

I first became aware of John in 1961 when I read his article "The Kayak of the Eskimo" (1961) in the journal *American White*

John Heath (left) and Kâlêraq Bech in Greenland, 1992. Courtesy Gail Ferris.

Water. We corresponded occasionally after that. *Sea Kayaker* magazine provided a valuable vehicle for knowledge about Native watercraft, spreading the message to modern recreational paddlers by publishing some eighteen articles and contributions from John, varying from how to make a Greenland-style paddle to a biography of Manasse Mathaeussen. Academia recognized his expertise by featuring his survey drawing of a two-hole *baidarka* from southern Alaska and description of the standard Greenland roll in the prestigious Smithsonian publication *The Bark Canoes and Skin Boats of North America* (1964). John followed with contributions to the Canadian Mercury Series and to the edited volume *Contributions to Kayak Studies* (1991). For some time, John had been working on a manuscript on traditional kayaks. It evolved into this book, which was at the University of Alaska Press at the time of his death. His wife Jessie and son David helped the editor with the final details.

Although I had known John for many years, I met him on only three occasions. These were informative, pleasurable, and memorable, and I held him in the highest regard. The traditional kayak community is the worse for his passing.

Duncan R. Winning, OBE
Honorary President, Scottish Canoe Association
Honorary President, Historic Canoe and Kayak Association
September 2003

Contents

	Map of the North American Arctic	*viii*
	Introduction	*ix*
	Acknowledgments	*x*
	Note on Inuit Terms	*xi*

PART I **GREENLAND**

	Maps of Greenland	**2**
John D. Heath	**1 — Kayaks of Greenland**	**5**
	Evolution and Construction	5
	Old Greenland Kayaks in Museums	8
	Greenland Kayak Paddles	14
	Origins of the Greenland Technique	15
	Training and Technique	19
	Capsize Maneuvers	20
	Storm Technique	31
	Sculling Rolls	33
	Additional Techniques	35
	Throwing Stick and Hand Rolls	38
	Miscellaneous Remarks	41
	The Greenland Kayak	42
	References	44
Greg Stamer	**2 — Using Greenland Paddles: An Overview**	**45**
Harvey Golden	**3 — Kayaks in European Museums: A Recent Research Expedition**	**61**

Hugh Collings	**4 — A Seventeenth-Century Kayak and the Swedish Kayak Tradition**	**75**
John Brand	**5 — Kayaks in England, Wales, and Denmark: Excerpts from *The Little Kayak Book* Series**	**83**
H. C. Petersen	**6 — Kayak Sports and Exercises**	**99**
Johannes Rosing	**7 — A Dramatic Kayak Trip, 1899–1900: Ataralaa's Narrative**	**103**

PART II **THE EAST CANADIAN ARCTIC**

	Maps of the East Canadian Arctic	**108**
E. Arima	**8 — Kayaks of the East Canadian Arctic**	**111**
	Introduction	111
	Kayak Hunting at Sea	112
	Kayak Hunting on Lakes and Rivers	118
	Variations	123
	The Multichine	123
	Flat Bottoms	124
	Kayak Making	128
	Speculations on Design Ancestry	137
	References	146

	Glossary	*149*
	Index	*153*
	About the Authors	*161*

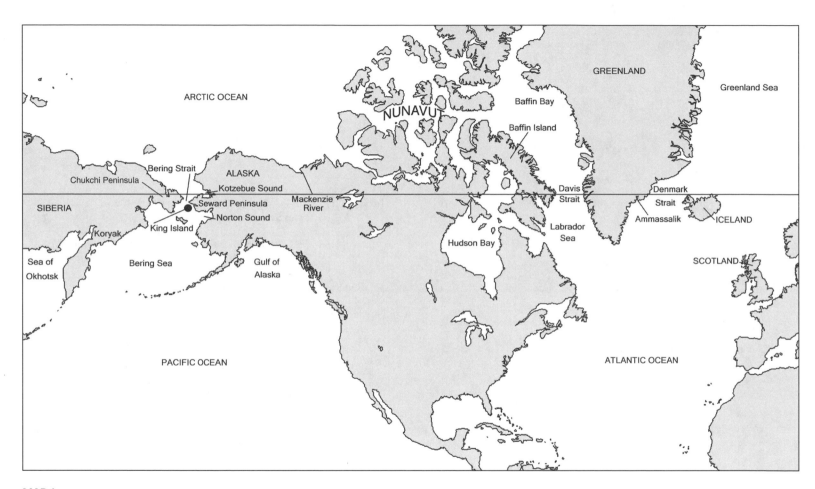

MAP 1

The North American Arctic and Greenland.

The bold black line indicates the location of the Arctic Circle.

Introduction

Arctic kayaks are ancient watercraft and are among the most complex of Native arctic technology. As working craft, they provide water transportation and enable people to hunt, trade, socialize, and wage war, raising the standard of living despite the costs of their laborious construction and maintenance.

Sea mammal hunting is the primary reason for the existence of kayaks; however, kayaks have also been used for caribou lancing, especially among the inland inhabitants of the Chukchi Peninsula, northern Alaska, and West Hudson Bay. Waterfowl and fish are taken using kayaks in many regions.

The present work is a comparative study of kayaks that focuses on historical development, design and construction, and techniques. For a study devoted to the traditional uses of kayaks, the reader is referred to H. C. Petersen's *Skinboats of Greenland* (1986). This outstanding work also covers the *umiak*, the open skin boat used in the Arctic. *Umiat* and kayaks are generally complementary craft. We have chosen to focus on the kayak in this volume in order to accommodate the interests of modern recreational kayakers.

Eastern Arctic Kayaks is divided into two geographic sections: Greenland and the East Canadian Arctic. John D. Heath, in the first chapter, places Greenlandic kayaks in historical perspective and describes rolling techniques. Heath's work is complemented by Vernon Doucette's invaluable sequential photographs. In Chapter 2, Greg Stamer contributes a detailed descriptive analysis of Greenland paddling. Recreational kayakers will find Stamer's work of great practical interest. Harvey Golden and Hugh Collings each offer chapters on kayaks in European and Greenlandic museums. Collings's chapter includes a discussion of the Skokloster Palace kayak — a key specimen for the study of Greenlandic design development. A treasure for North American readers is Chapter 5, an excerpt from *The Little Kayak Book* series by John Brand, which presents data on a number of significant kayaks in British collections accompanied by exquisite lines drawings.

The final two chapters in the Greenland section were written by Greenlanders and offer Native perspectives on kayak use. H. C. Petersen describes how young kayakers gain expertise with watercraft; Johannes Rosing recounts a kayaking ordeal during a storm at the turn of the last century.

The second half of this volume is devoted to kayaks of eastern Canada. Included in the chapter by E. Arima are historic photos and excerpts from oral history interviews describing hunting by kayak in North Baffin and the Foxe Basin.

The kayak design is an ancient one. Archaeological evidence indicates that kayaks have been in use for at least four thousand years, although the record of kayak development has remained sketchy until the last few centuries. Historical accounts indicate that as many as fifty or sixty kayak types can be identified. These "types" can be grouped into a dozen or so "families." *Eastern Arctic Kayaks* is about one such family.

Kayaks are a unique arctic achievement, a fascinating aspect of northern cultural heritage. When out paddling in the wind or calm, on water salt or fresh, we might remember to thank the arctic hunters who developed and refined the exquisite craft we call "kayak."

E. Arima

Acknowledgments

John Heath: Many individuals and organizations provided assistance in gathering the data for this volume. By letting me try out their folding kayaks in 1954, Ward and Lois Irwin helped rejuvenate a childhood interest in kayaks. Admiral MacMillan brought awareness of what Greenlanders can do. Howard Chapelle taught me how to take off the lines of kayaks. I am also grateful to individuals in the following organizations: National Historic Parks and Sites of Canada, United States National Museum (Smithsonian Institution), American Museum of Natural History, National Museum of the American Indian, Mariners' Museum (Newport News), Canadian Museum of Civilization, Royal Ontario Museum, Field Museum, Milwaukee Public Museum, Missouri State Historical Society, British Museum, National Maritime Museum (Greenwich, U.K.), Royal Scottish Museum, Surgeons' Hall (Edinburgh), Marischal Museum (Aberdeen), the Aberdeen Medico-Chirurgical Society, Kelvin Hall Museum (Glasgow), Hunterian Museum (Glasgow), University of Dundee Museum, Perth Museum and Art Gallery (Perth, Scotland), the National Museum (Copenhagen), the Greenland National Museum (Nuuk), Maniitsoq Museum (Illunnguit), Sisimiut Museum, Qasigiannguit Museum, and Knud Rasmussen Museum (Ilulissat, Greenland).

Institutions not personally visited but providing specimens presented in the volume include Newfoundland Museum, Trinity House in Hull, University of Cambridge Museum of Archaeology and Anthropology, Braintree Town Hall, Lancing College, Atlantic College, Skokloster Palace, and the Ethnographic Museum in Stockholm.

Helpful to me in Greenland in many ways have been the staff of Knud Rasmussen High School and the members of the Greenland Kayaking Association (Qaannat Kattuffiat), among whom Kâlêraq and Lone Bech have been my frequent contacts, but if space permitted dozens of others could be named. First Air and Greenland Air have provided assistance to the Greenlanders who brought their kayaks to the U.S. on several occasions for kayaking exhibitions. By steadfastly supporting the Greenland Kayaking Association, the Royal Greenland A/S organization has indirectly helped to make this book possible.

I also wish to thank the following individuals and organizations who generously gave permission to include their photographs in this publication: Keld Hansen and Ove Bak of Denmark, Henning Redlich of Germany, Jens Ostergaard, Vernon Doucette, the Danish Arctic Institute, and the Library of Congress Prints and Photographs Division. Their contributions are greatly appreciated.

Every effort has been made to properly spell and use Greenlandic terms and place names. In some cases several spellings for a single location were identified. For example, Nuuk, the capital of Greenland, has been variously spelled Nuuk, Nûk, Godthåb, and Godthaab. To further complicate the issue, there are several dialects within Greenland. If, in spite of our best efforts, spelling errors have crept into the text, we ask for the reader's forbearance.

E. Arima: Many of the fine Greenland kayak photographs in this volume are due to the generosity of Helge Larsen, Keld Hansen, and Ove Bak of Denmark, and Gert Nooter of Holland. Several photos are published courtesy of the National Archives of Canada, the Canadian Museum of Civilization, and the Scott Polar Research Institute.

Note on Inuit Terms

SPELLING AND PRONUNCIATION

The word "kayak" is the English equivalent for a word spoken by Native Inuit. The word is made with the tongue held against the uvula to produce a "q" sound (instead of a "k"). The "a" sounds are pronounced like the "a" in "father." The word for Inuit watercraft should thus be spelled "qayaq" in English, or "qajaq" in Danish or German. We have yielded in this volume to popular usage of the term "kayak."

The syllabic system is the most common Canadian Inuit writing system now in use, with some regional exceptions. A three-vowel system is usually followed; **u** ranges from "o" to "oo" and **i** from "ee" to "e" as in "heck," while **a** is usually pronounced like "ah."

Vowel lengthening is shown by doubling the letter, i.e., **aa, ii, uu**. Alternative vowels will appear at times for terms from Greenland where **o** is still in frequent use. The **q** and **r** are pronounced in the back of the throat; the Inuit **k** and **g** sounds originate farther back than they do in English. It may be helpful to liken the difference between **k** and **q** to that between the consonants in German "ich" and "ach." For those consonants made farther back in the throat, vowels are naturally sounded more openly; that is, the **u** is pronounced as it is in "oh," and **i** as in "pick." Vowel lengthening can also be indicated by a macron bar or circumflex, a practice used in Greenland. Consonant clusters are variously written. Thus, bowhead whale might be either "aqviq" or "arviq," and bearded seal "udjuk," "ugjuk," or "ujjuk." Similarly, house could be written "illu," "idlu," "iglu" and, of course, "igloo."

PLACE NAMES

Another complication is that Inuit place names are replacing many European ones on maps. Mittimatalik (northeast Baffin Island) used to be labeled Pond Inlet. Spellings change, too. "Mitimatalik" was spelled with only one "t" when it was commonly, but incorrectly, believed to mean "where geese land."

Figures in this volume are labelled by locality, and occasionally by the name of the associated Native group. This approach can pose problems. Thus, although the people of northwest Greenland call themselves Inuhuit, "Great Inuit" (Gilberg 1984:593), it is practical to use "Polar Inuit" to distinguish their kayaks outside Greenland. Greenland is now Kalaallit Nunaat, or "Land of the Kalaallit." "Greenland kayak," which is frequently used in this book, is so well established that switching to "Kalaallit" may be confusing.

If kayak study becomes more of an Inuit concern, Native group names may eventually predominate in specimen designations. The line drawings in this volume preserve the various designations that have been used over the years; once applied, a label is likely to stick as the identifier of the specimen, especially after the drawing is published or otherwise circulated.

PART I
Greenland

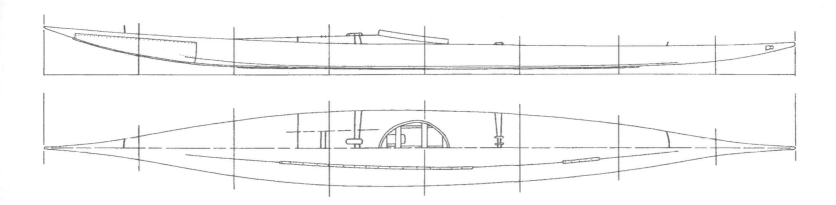

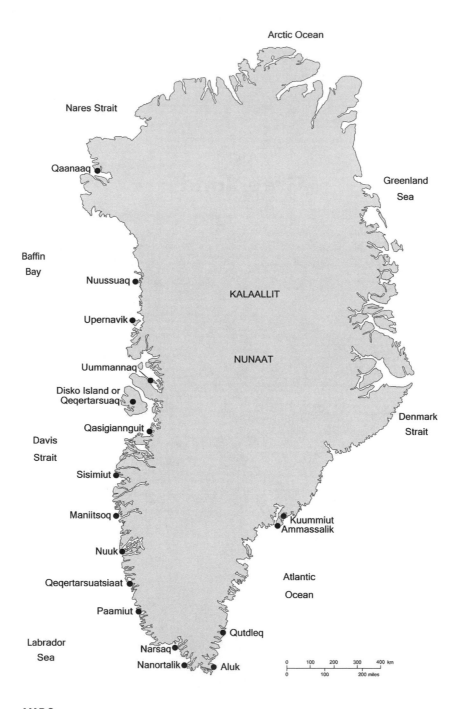

Arctic Ocean

Nares Strait

Qaanaaq

Greenland
Sea

Baffin
Bay

KALAALLIT

Nuussuaq

Upernavik

NUNAAT

Uummannaq

Disko Island or
Qeqertarsuaq

Qasigiannguit

Denmark
Strait

Davis

Strait

Sisimiut

Maniitsoq

Kuummiut
Ammassalik

Nuuk

Qeqertarsuatsiaat

Atlantic

Ocean

Paamiut

Qutdleq

Labrador

Sea

Narsaq

Nanortalik   Aluk

0   100   200   300   400 km

0   100   200 miles

MAP 2

Kalaallit Nunaat (Greenland).

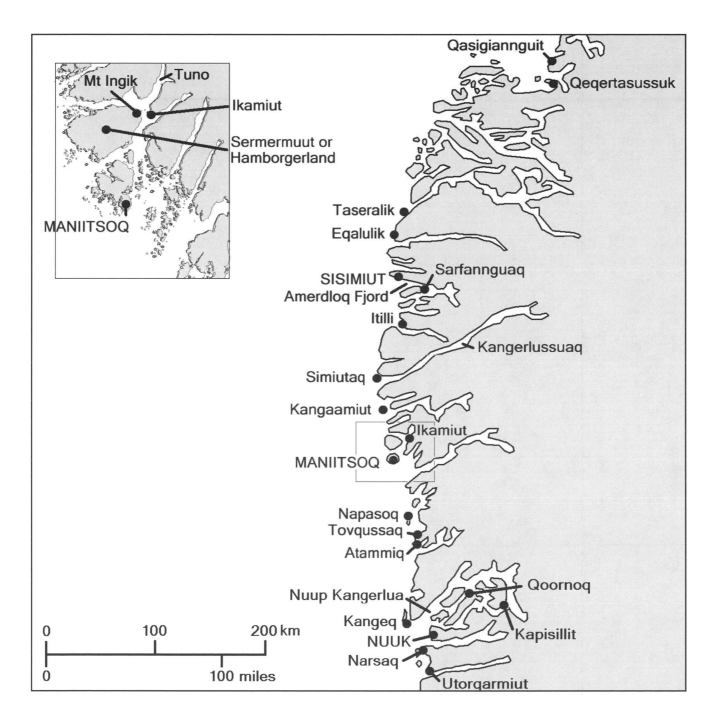

Qasigiannguit

Qeqertasussuk

Mt Ingik   Tuno

Ikamiut

Sermermuut or
Hamborgerland

MANIITSOQ

Taseralik

Eqalulik

SISIMIUT   Sarfannguaq

Amerdloq Fjord

Itilli

Kangerlussuaq

Simiutaq

Kangaamiut

Ikamiut

MANIITSOQ

Napasoq

Tovqussaq

Atammiq

Qoornoq

Nuup Kangerlua

Kangeq

Kapisillit

NUUK

Narsaq

Utorqarmiut

0   100   200 km

0   100 miles

MAP 3
Western Greenland.

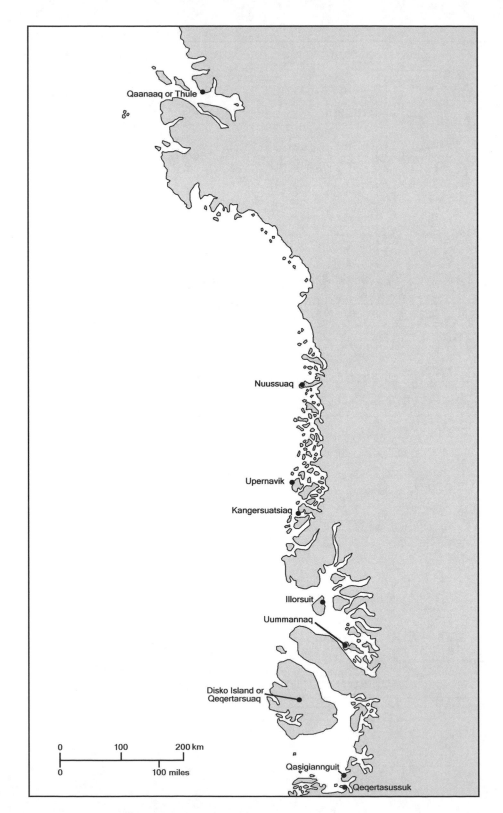

Qaanaaq or Thule

Nuussuaq

Upernavik

Kangersuatsiaq

Illorsuit

Uummannaq

Disko Island or
Qeqertarsuaq

0   100   200 km

0   100 miles

Qasigiannguit

Qeqertasussuk

MAP 4
Northwestern Greenland.

1 Kayaks of Greenland

JOHN D. HEATH

This chapter will discuss the origin and antiquity of the Greenland kayak. The natives of each arctic region where kayaks were used developed and adapted the design for their needs. When one travels from west to east or vice versa along the full range of the kayak, one can almost see an evolution progressing. I will point out some important clues hidden in that design transition.

Modern kayak enthusiasts want more and more information on how Greenland kayakers used their crafts. I will explain the distinctive Greenland paddle and why it needs to be the way it is. I will then devote the bulk of my chapter to the how and why of the "Eskimo roll." Several Inuit peoples knew how to right an upside-down kayak by ingenious methods. However, I feel that the Greenlander took this life-or-death maneuver to its highest and most diverse perfection.

I will close with some brief notes and a paraphrase of an excellent speech by Greenlander Kâlêraq Bech, kayak enthusiast and past president of the Greenland Kayaking Association (Qaannat Kattuffiat).

EVOLUTION AND CONSTRUCTION

For many years, two kayaks have been stored side by side at the American Museum of Natural History in New York. I observed the two specimens in 1963, 1969, and 1988. How odd that out of the many kayaks in the museum, these particular specimens happened to be placed together—two kayaks from the opposite outer fringes of the region where kayaks are known to have been used. One is a Koryak kayak (Figure 1.1) (acc. no. 70/3358) from the Sea of Okhotsk region of eastern Siberia (Map 1). The other is an East Greenland kayak (60.1/6003) from the Ammassalik District (Map 2), which is similar to the kayak depicted in Figure 1.2. The two regions are at opposite ends of the migration routes used by ancient northern peoples as they hunted sea mammals in Siberia, North America, and Greenland.

The Koryak kayak is less than half as long yet has almost half again as much beam as the East Greenland kayak. The Koryak kayak is propelled by two short paddles, one held in each hand. The paddles are attached to the cockpit hoop by lanyards. The East Greenland kayak is propelled by a narrow, double-bladed paddle. Seeing the two kayaks side by side tends to underscore the differences between them. Yet some striking similarities in Koryak and East Greenland kayaks were observed in these American Museum of Natural History specimens and cited in a comparative study:

1. Both have a "floating" cockpit rim or hoop (see glossary) that is attached to the deck by the skin only.
2. Both have an essentially flat deck with two or more longitudinal stringers in the deck.
3. Both have three longitudinals attached to the ribs, consisting of a central keelson and two bilge stringers.

The presence of these common features at the opposite outer fringes of Inuit habitation could indicate that there was an ancient archetype from which all kayaks evolved, or it could be entirely coincidental. These features deserve further investigation. If there are any ties between the kayaks of the Sea of Okhotsk and those of East Greenland, then the probable path of any cultural impulses connecting the two regions would have been across the Bering Strait and along the northern coast of the Seward Peninsula. From

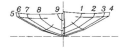

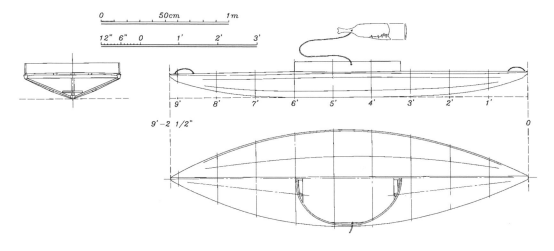

Figure 1.1
KORYAK KAYAK
American Museum of Natural History, no. 70/3358
Collected by the Jesup Expedition

Length	9' 2 ½"	280.7 cm
Beam	29 ½"	74.9 cm
Depth	8"	20.3 cm

HARPOON REST (STARBOARD ONLY)

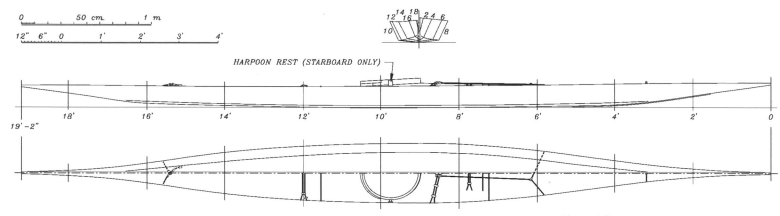

Figure 1.2
EAST GREENLAND KAYAK
Greenland National Museum, no. KNK 525

Length	19' 2"	584 cm
Beam	18 ½"	47.0 cm
Depth	6 ½"	16.5 cm

the Mackenzie River region eastward to Greenland, all kayaks had the "floating" cockpit hoop that was also found in Siberia. Yet in Arctic Alaska, from Kotzebue Sound northward, the cockpit hoops are an integral part of the framework of the kayak. This feature evidently made its way northward from the Bering Sea (Heath 1978:21–22).

The Koryak and East Greenland specimens compared above were part of a total of one hundred kayaks examined in 1963. These included the kayak collections of the United States National Museum (now the Smithsonian Institution); the American Museum of Natural History, New York; the Museum of the American Indian (formerly in New York City, now the National Museum of the American Indian, Washington, D.C.); the Canadian Museum of Civilization, Gatineau; and the Royal Ontario Museum, Toronto. Examining a diverse assortment of kayaks within a relatively short period of time was helpful in this particular comparative study; however, evidence on form and construction must be supported by other data. Any theories about a connection between Koryak and East Greenland kayaks must have supporting archaeological data in order to be conclusive.

The kayaks of northern Alaska, Canada, and Greenland share several common features. For example, from Kotzebue Sound, Alaska, eastward to Greenland, all kayaks had raked cockpits—that is, the cockpit rims are highest at the foremost edge and lowest at the aftermost edge. This facilitates entry and exit, provides legroom under the foredeck, helps keep water out of the cockpit when the kayak was under way, and provides elbow clearance when paddling. Another feature of all kayaks from Kotzebue Sound eastward is the use of deep gunwales to provide hull rigidity. When a skin-covered craft is under way, the hull sags to follow the water surface unless it is rigid enough to bridge across the wave crests. Or, if the ends of the kayak project out of a wave crest, the ends droop and the midsection is lifted by the wave, an action that is called hogging. A kayak that is too flexible will tend to hog and sag alternately in rough seas, making it difficult to control. Deep gunwales are a simple way to make a flat, shallow kayak rigid enough, yet not too rigid. South of the Bering Strait, hull rigidity is achieved by having a broad, deep hull. Koryak kayaks are shorter and are not used in rough seas, so deep gunwales or deep hulls are unnecessary.

The lineage of Greenland kayaks can easily be traced westward as far as Kotzebue Sound, just north of the Bering Strait. The deep

gunwales, flat afterdeck with raised foredeck, and raked cockpit are apparent at a glance. A closer examination shows that the longitudinal stringers are lashed to the ribs in such a way that the lashing runs lengthwise along the stringers, a feature of kayaks north and east of the Bering Strait. All Alaskan kayaks south of the Bering Strait have this lashing running crosswise along the ribs. This large area of similar kayak features, found from Kotzebue to Labrador, was also the range of the Thule culture.

Kayaks of Greenland and eastern Canada have a uniquely deep forefoot. That is, the bow or forefoot of the kayak has an exaggerated vertical dimension (see glossary diagram on page 150). The eastern part of the Thule range is the only place where *both* Thule and Dorset culture sites are found. That the deep forefoot is confined to that area may suggest that it might have been a feature of Dorset kayaks that was adopted by the Thule culture as they migrated eastward. Yet there is no hard evidence that the Dorset culture even had kayaks. From available archaeological data, there is only a suggestion that the Dorset culture had some type of watercraft.

Trees large enough to construct this type of kayak do not grow in the region. However, large-enough logs occasionally wash up on the beach. The gunwales are split from a driftwood log and shaped together so that they match. Holes for receiving the ends of the thwarts (cross beams or deck beams) and ribs are made in the sides and bottom edges of the gunwales. Then the upper framework, consisting of the gunwales and thwarts, is assembled. Next, the ribs are bent into shape and each end of a rib is inserted into holes along the bottom of the gunwales, starting in the middle and working toward each end. Bilge stringers and the keelson can be laid along the ribs as the latter are installed, to make sure that they are properly aligned. If not, the ribs can be taken between the teeth and bitten to work more bend into them at critical places.

Greenland kayaks used to be made in much the same manner as the kayaks from northern Alaska to eastern Canada, that is, the stem and stern bottom pieces were essentially an extension of the keelson. There are a number of old kayaks in Europe dating from the 1600s and 1700s that feature variations on this construction (see chapters by Harvey Golden and Hugh Collings, this volume).

When all of the ribs are in place, the keelson is attached (Figure 1.3). This is sometimes in one piece, but construction varies locally; in most places, the keelson is made in two or three pieces, consisting of plank-on-edge end pieces that are joined by a scarf in the middle and nailed, lashed, or pegged together. The keelson is attached to

Figure 1.3—Northwest Greenland kayak framework, Qaanaaq. The gunwales are constructed of several pieces of wood scarfed together. © Jens Ostergaard, 1995.

If two otherwise waterproof skins are sewn together by sewing completely through a skin, then the skin will leak at every puncture. This is because the side force on the thread will elongate the hole through the skin. Typically, underwater seams in kayaks are made by overlapping the pieces of skin about an inch (2 or 3 cm) and, using a curved needle and a running stitch, sewing through the edge of the top skin, but only halfway through the lower skin. Then the process can be repeated from the opposite side. This means that no stitching hole passes entirely through both skins. Assuming that the skins are horizontal, there are no vertical holes through both skins caused by the sewing. Thus, the only place for leakage is horizontally through the two or three centimeters of overlap. This interface is greased during the sewing process, which makes the kayak waterproof. However, if the kayak remains wet too long, it will eventually become waterlogged. Kayakers try to allow their craft to dry out every night and keep the skins greased (Figure 1.4).

During the application of the skin covering, the lengthwise deck seams are sewn by first pulling the skins together with a large zigzag lacing stitch somewhat like lacing up a shoe. This pulls the skin to a tight fit around the framework. The lacing is sewn partway through from underneath, to avoid causing leaks. The tension of this lacing causes the puckers visible on the deck of the kayak at the places where it pulls on the skin. Then the skin is sewn shut the length of the deck from each end to the opening at the cockpit. The cockpit hoop is attached by either pegging it or lacing it to the deck skin. Deck straps are attached to hold the implements, with the foremost and aftmost straps installed as the deck seam is sewn.

The description above is a brief, general sketch of kayak construction. More detailed data can be found in the works of H. C. Petersen (1982, 1986). Local Greenland methods varied considerably, yet the Greenland kayaks were much more standardized than those in other parts of the region where kayaks were used.

OLD GREENLAND KAYAKS IN MUSEUMS

Most Greenland kayak museum specimens were collected after about 1850. Thus, our knowledge of earlier Greenland kayaks is based mainly on a few old kayaks in the museums of northern Europe (Brand 1984, part I:2–10; Gabler 1982; Nooter 1971; Reid 1912:511–514; Souter 1934).

In 1971, the fragments of a kayak framework were found at Eqalulik, 22 miles (35 km) north of Sisimiut (Map 3) on the west coast of Greenland. This material was brought to H. C. Petersen,

the ribs and ends of the gunwales by lashing, as a rule, but in some localities pegging or nailing is used.

The two side stringers are then aligned visually and lashed into place. Most West Greenland kayaks have two short deck stringers in front of the cockpit and two aft of the cockpit, to support the deck skin where it has the greatest span. The cockpit hoop on all Greenland kayaks is attached to the skin only; in other words, it floats.

The skin of West Greenland kayaks is usually in four pieces; that is, it has three transverse seams, but it may be covered with only two sealskins, one in the middle and the other cut to make the end pieces, so that there are a total of only two transverse seams. The skins are usually sewn together and fitted on the framework, wrapping around the bottom and hooking onto the ends with two small thimble-like pockets of skin made wrong side out and arranged to hook over the stem and stern.

Figure 1.4—Kayaks at Kangersuatsiaq, Upernavik region, West Greenland, 1881. The kayaks are stored out of reach of dogs. The float is tied over the cockpit. Photographer Sgt. George W. Rice of the U.S. Army Signal Corps. Courtesy Library of Congress, LC-USZ62-38037.

then headmaster at Knud Rasmussen High School in Sisimiut, who arranged the pieces and exhibited them at the Sisimiut Museum (Petersen 1986:56–57). The Eqalulik kayak is especially useful for research because the pieces are arranged so that they can be easily examined. The kayak paddle and other implements that were found with the kayak are also on exhibit at the Sisimiut Museum.

The kayak fragments indicate that the construction was similar to that of several old kayaks in Holland (Nooter 1971). Although the Eqalulik kayak had the plank-on-edge gunwales and bent rib construction of later Greenland kayaks, there are several differences in construction. The gunwales lack the end pieces of later kayaks and extend to the ends of the kayak. In this respect they are similar to 19th- and 20th-century specimens from northern Alaska and most Canadian kayaks. The thwarts were set at different heights, as has been observed in other old Greenland kayaks (Petersen 1986:56). The ends of the thwarts rest in holes in the gunwales. But these holes do not go all the way through, so the lashing that holds the thwarts in place runs across the kayak in order to secure the thwart in position. Later kayaks have a different means of securing the thwart in position. They used a vertical hole made in the thwart near the gunwale and a diagonal lashing from this hole to one in the lower edge of the gunwale.

In 1992, I met with H. C. Petersen at the Sisimiut Museum, where it was possible to examine and discuss the Eqalulik kayak. This old kayak had many of the features found on 17th- and 18th-century kayaks in other collections. Petersen had already estimated the age of this specimen: "Taken together, the evidence indicates that the Eqalulik

kayak stems from either the end of the 17th or the first half of the 18th century" (Petersen 1986:57). In 1995, I reexamined the Eqalulik specimen and found no reason to differ with Petersen's estimate.

There are other old kayak specimens in northern European museums that furnish data on the shape and construction of 17th- and 18th-century kayaks. In general, the old kayaks tend to be longer and narrower than the later ones. Most of them taper toward the ends with convex lines, as seen from above, instead of the concave ends that are typical of later Greenland kayaks. Another difference is that the cockpits of the old kayaks tend to be located closer to the lengthwise center of the kayak, as in Figure 1.5. More modern kayaks generally have the cockpits slightly aft of center.

Gert Nooter, of the Rijksmuseum voor Volkenkunde in Leiden, has done much to help researchers understand the evolution of kayaks in Greenland (Nooter 1971). In 1983, the Rijksmuseum loaned three old kayaks to the Greenland National Museum for a special exhibit. This exhibit came at a time when kayaking had significantly declined in West Greenland. Thimothaeus Poulsen, one of the founders of the Qajaq Club at Nuuk, told me that the only knowledge many West Greenlanders had of kayaks was from seeing the performances of Manasse Mathaeussen when he toured Greenland demonstrating kayaking technique. Nooter's loan of the old kayaks from Dutch museums gave Greenlanders a chance to see that kayaks had evolved into sleek vessels centuries ago and instilled a sense of pride in those who saw it.

Figure 1.5
CENTRAL WEST GREENLAND KAYAK, c. 1700
Marischal Museum, ABDUA:6.013
University of Aberdeen, Aberdeen, Scotland
Surveyed 1986; drawn to correct for damage

Length	17' 11½"	547.4 cm
Beam	17¾"	45.1 cm
Depth to sheer	7"	17.8 cm
Depth, foredeck	8"	20.3 cm

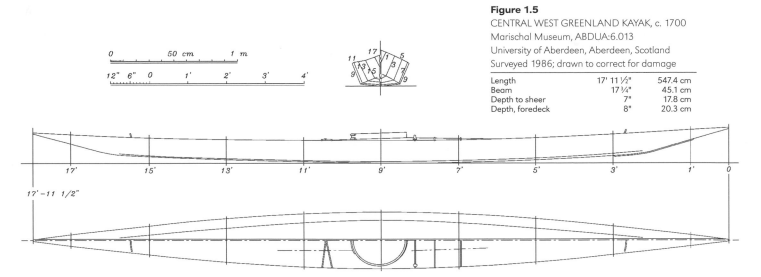

One of these kayaks was given to the Greenland National Museum (cat. no. KNK 1161), and it was this kayak (Nooter 1971:figs. 26, 27) that I examined at Nuuk in 1985 and again in 1992. This example is typical of several other old kayaks from the central West Greenland coast in that it is long and narrow. However, the shape, as seen from above, lacks the concave appearance at the ends. As seen in profile, the sheer line does not curve upward very much and the stem and stern are rather short. One of the striking features of this kayak is that the clear space available between the deck beams that support the forward and after edge of the cockpit hoop measures only 10½ inches (26.7 cm). Since this dimension limits the space through which the kayaker must wriggle into and out of the kayak, the kayaker must have been small and limber by modern standards. Yet this appears to have been a full-size kayak in all other respects.

One old kayak that was probably made for a child is a specimen in Trinity House, Hull, England (see chapter by Brand, this volume). Charles Ranshaw and John Brand surveyed this specimen in 1966 and Brand made drawings, which are reproduced on page 84). The length of this specimen is 11 feet 10¼ inches (361.3 cm). It has a beam of 1 foot, 5 inches (43 cm). These are consistent with the dimensions of a child's kayak. The lines of the kayak are similar to those of more modern kayaks from the vicinity of Nuuk, West Greenland. This specimen may have been collected by the James Hall Expedition of 1612, since there was a reference to kayakers paddling with the two ships of that expedition on August 13, 1612. The location was 64 degrees north latitude on the West Greenland coast, near the mouth of Nuup Kangerlua, formerly Godthaabs Fjord (Gosch 1897:111).

Another kayak dated 1607 hangs in the Schiffergesellschaft at Lübeck, Germany (Gabler 1982) (Figure 1.6). This specimen could have been collected on the Danish Arctic Expeditions of 1605 or 1606, because each of these expeditions acquired kayaks in West Greenland. However, the Danish Arctic Expedition of 1607 did not land in Greenland because of ice conditions (Gosch 1897). So the date "1607" painted on the kayak probably applies to the date it arrived in Lübeck, possibly by way of Denmark, as a gift, or by capture of a Greenlander who fled from Danish custody. Although the Lübeck kayak has an earlier date than the Trinity House specimen, the two craft are only a few years apart, so it is impossible to be certain which is actually older.

Another old kayak from the 17th century is at the Wrangel Armoury, Skokloster Palace, Sweden. This specimen was recently surveyed and drawn up by Hugh Collings (this volume). It appears to be the best-preserved 17th-century kayak in existence, which speaks well of the museum personnel who took care of it for more than three centuries. An unusual feature of this kayak is the half-round shaped lip carved as an integral part of the wooden cockpit hoop. This is found on some kayaks from eastern Canada, but it is an extremely rare feature on a Greenland kayak.

I have examined some old kayaks in the United Kingdom that have a gracefully curving upturned stern and some that have the same type of shape at the bow. These seem to have been used in some areas of West Greenland around 1800. The shape is similar to that of the stern of a Caribou Eskimo kayak, except the latter has an abrupt upturn that then becomes straight for most of its length. The old Greenland-style stern has a more gradual curve. The kayaks that have the high, upwardly curved bow and stern are almost symmetrical at the bow and stern, but the stern is usually slightly longer and higher. The kayaks without the high curved bow are slightly deeper at the forefoot.

A kayak at the Aberdeen Medico-Chirurgical Society, Scotland, displays the old-style Greenland stern. William Souter measured the kayak and published it (Souter 1934). In 1986, I examined this kayak, which was still at the Medico-Chirurgical Society and was in good condition. A similar kayak resides at Surgeons' Hall in Edinburgh, Scotland. The Edinburgh kayak has the high, curved stern that is typical of the type, but it was exhibited high atop a display cabinet and was inaccessible for close examination.

A noteworthy old kayak with a well-documented date can be found at the Hunterian Museum in Glasgow, Scotland. This kayak (E102) has the high, gradually curved stern of the kayaks described above and is in excellent condition. Museum records are precise about the date of collection: October 27, 1790. This helps establish the period in which this style of kayak was used. Unfortunately, no record was found that indicated exactly where in Greenland it was collected.

Two other old Greenland kayaks at the Hunterian are also well dated: both date to 1819. One has the lines of later Greenland kayaks and one has the high, upturned ends at bow and stern. The date of 1819 is helpful in establishing the period during which the kayaks with the long, graceful, upturned ends were made in Greenland.

In the Marischal Museum at the University of Aberdeen is specimen no. V-87, a beautiful kayak with upturned ends that was collected in Greenland in 1800. It has alternating light and dark skins, which Greenlanders have used for centuries to create an attractive kayak. Unfortunately, this otherwise pristine specimen

Figure 1.6a

1606 GREENLAND KAYAK FRAMEWORK

Schiffergesellschaft, Lübeck, Germany

Adapted from Gabler (1982:38)

Length	17' 4⅝"	529.9 cm
Beam	19¼"	48.9 cm
Depth to sheer	7½"	19.1 cm

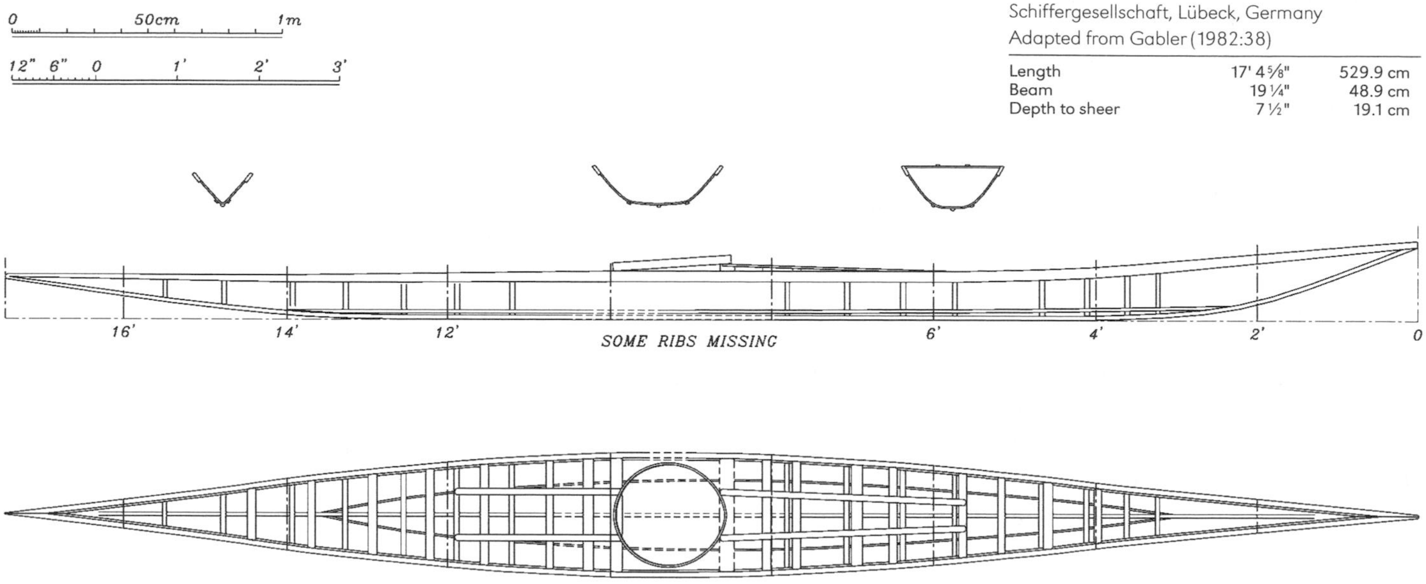

SOME RIBS MISSING

Figure 1.6b—Forward interior of the Lübeck kayak.
© Henning Redlich.

Figure 1.6c—Aft interior of the Lübeck kayak.
© Henning Redlich.

was sawed in half, perhaps to make it fit in the ship that brought it from Greenland. Only a piece of skin was left intact at the side to keep the two halves joined.

Another Marischal Museum specimen, no. ABDUA:6013 (Figure 1.5)—perhaps the most famous kayak in the world—has intrigued researchers for centuries. The unusual story of how this kayak arrived in Aberdeen is found in the 1824 List of Contents of the museum:

> Esquimaux Canoe, in which a native of that country was driven ashore near Belhelvie, about the beginning of the eighteenth Century, and died soon after landing.

The List of Contents is in a hardbound journal. Curiously, at the other end of the same journal is another entry, from Archive no. M-106, the Inventory of Marischal Museum: "Esquimaux Canoe, in which a native is said to have been driven ashore here about 1700—."

Both entries are vague about the exact date, but one mentions the approximate location where the kayak was driven ashore, and adds that the kayaker died soon after landing. Belhelvie, the town near which the kayaker was driven ashore, is just north of Aberdeen. Both journals refer to the kayaker being driven ashore, but it is unclear whether sea conditions or people were responsible.

There has been a lot of speculation as to how a kayaker arrived in Scotland under his own power at the beginning of the 18th century. The mystery was deepened by the reports of "Finnmen" in the Orkneys (MacRitchie 1912:493–510; Wallace 1700:60–61). But the only hard evidence, the Marischal Museum kayak itself, was identified as being a typical kayak from central West Greenland (Birket-Smith 1924:266). In 1934, Souter suggested that the Aberdeen kayaker escaped from a whaling ship that was returning from the Arctic and made it to shore (Souter 1934:17).

In the 1950s, I read the published account of the Marischal Museum specimen, discussed above, as well as those of Mikkelsen (1954) and Whitaker (1954). In 1951, an article about Greenland appeared in the October *National Geographic*. Miriam MacMillan, the wife of arctic explorer Admiral Donald MacMillan, wrote about dropping a bottle with a note in it off the coast of West Greenland, near the Arctic Circle. Two months later, the bottle was returned by a Scotsman, who had found it in the Orkneys.

In 1954, Ian Whitaker proposed that the Aberdeen kayaker paddled from Greenland to Scotland by way of Iceland and the Faeroe Islands. The same year Whitaker made his proposal, Ejnar Mikkelsen suggested the kayaker escaped from a whaler (Mikkelsen 1954:5, 7). This led me to hypothesize how a West Greenland kayaker happened to arrive in Scotland under his own power.

The coastal current off West Greenland flows northward. Thus, if a West Greenland kayaker were blown to sea or lost in fog and survived the night, he would know how to get home. He would have to paddle eastward until he reached the coast and then southward. But if he were at sea for several nights, he might land on an ice floe and survive by eating whatever birds, fish, or sea mammals he could capture with his hunting implements, some of which survive with the kayak today. Sea ice provides potable water on the upper surface and it would give the kayaker a chance to dry the kayak and even grease the skin with blubber from sea mammals.

Although the kayaker would be familiar with Greenland coastal currents, he would not know that on the Baffin Island side of Davis Strait the current flows southward. If he were at sea several days, he might get caught in part of the southward-flowing current and be taken southward. If the fog or storm persisted long enough, the kayaker could be drifting far offshore in a southerly direction, so that when conditions improved and he left his ice floe, he would be far south of where he perceived himself to be. Thus, when he started to paddle eastward, he would miss Greenland entirely. Perhaps he would doggedly keep paddling eastward if he could not see land in any direction, because his experience had taught him to head east in order to reach land, then head south to reach home. After he had started eastward in the North Atlantic, the prevailing wind and currents would have aided him.

I wrote a summary of this hypothesis in 1959, and sent a copy of it to Kaj Birket-Smith in 1962. In his reply, Birket-Smith was skeptical, and suggested that either the story about a Greenlander being picked up in Scotland is a myth or, if the Greenlander was picked up, he was trying to escape from a European ship (Birket-Smith, letter of March 24, 1962).

I used my unpublished 1959 report to prepare a more detailed article about the Orkney kayak sightings and the incident at Aberdeen (Heath 1987). Although it is easier to accept the hypothesis that the kayaker left a ship near Scotland or escaped from captivity in Denmark or Holland and made it as far as Scotland, I believe it was possible for the kayaker to have made it from Greenland to Scotland. In 1985, I expressed my opinion to two veteran seal catchers, Pele Holm, then

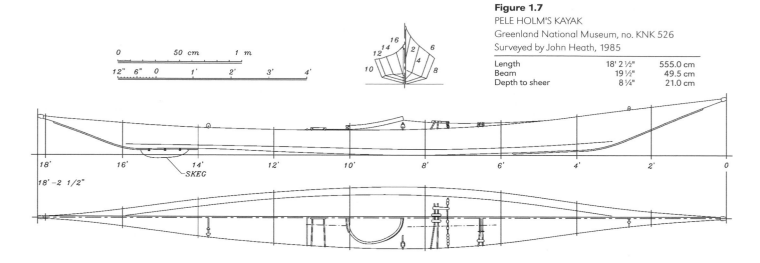

Figure 1.7
PELE HOLM'S KAYAK
Greenland National Museum, no. KNK 526
Surveyed by John Heath, 1985

Length	18' 2 ½"	555.0 cm
Beam	19 ½"	49.5 cm
Depth to sheer	8 ¼"	21.0 cm

85 years old, and Manasse Mathaeussen, then 70 years old. Both were skeptical at first but later agreed that it was possible. Mathaeussen thought the trip could be done in two weeks, if the kayaker was paddling with favorable storm winds. Mathaeussen was qualified to comment on the physical limitations of being at sea for long periods of time, as he was in the habit of lying on the surface of the water and taking naps on long trips. (See the description of the maneuver *nalaasaarneq*, "to lie down," in the section on technique.)

GREENLAND KAYAK PADDLES

Greenland paddles and kayaks are made to fit the kayaker by using anthropometric measurements. Various lengths may be described as follows:

1. For a short "storm paddle," a popular length is one arm span (the distance from fingertips to fingertips with arms extended).
2. For a general-purpose paddle to be used often for rolling maneuvers, a popular length is one arm span plus the distance from the wrist joint to the fingertips.
3. For a general-purpose paddle, a popular choice is one arm span plus the distance from inside the bent elbow to the fingertips.
4. For a general-purpose paddle, another choice is one arm span plus the distance from outside the bent elbow to the fingertips.
5. For a general-purpose paddle to be used by a seal catcher for towing seals alongside the kayak, a popular choice is one arm span plus the distance from the armpit to the fingertips.

Greenlander Pele Holm had a paddle that was 8 feet (244 cm) long. Most paddles that I have measured seem to fall between the second and fourth listed choice of lengths; however, there are no hard and fast rules. Pele Holm's kayak is shown in Figure 1.7.

Many people ask why Greenland paddles are so narrow. The almost universal assumption is that the paddles are narrow because Greenlanders do not have enough wood to make wider ones. Actually, the primary reason for the narrow paddle is so that it can be gripped firmly at any point along its length. This is confirmed by the anthropometric measurement for the width of the paddle blade. The width of the widest part of the blade is determined by making a **C** with the thumb and forefinger. The width is the distance between the web of the thumb and the second joint of the forefinger, that is, the maximum distance one can comfortably grip. The loom, or shaft, of the paddle (see the glossary diagram on page 151) has a circumference that is determined by touching the tip of the index finger to the tip of the thumb. Many kayakers increase this circumference to suit their personal preference. The loom is usually thicker at right angles to the blades so that it will be stronger in the direction that the paddling force is applied. This also allows a kayaker to know which way the paddle is rotated without looking at it. The length of the loom is slightly more than the width of the kayaker's hips.

The paddles are edged and tipped with whalebone in order to prevent the edges and the tip of the blade from fraying when paddling in ice-filled water. If the edges fray, the paddle will pick up water at each stroke, which drips and makes noise that a seal might hear, according to Pele Holm. The paddle tip is made from the jawbone of a whale and the bone edging is made from a whale rib.

The bone edging has another practical use. According to Jan Schmidt, a veteran seal catcher, when he entered his kayak, he would lay one end of his paddle across the kayak with the paddle on edge, so the other edge lay on the shore. This put his paddle edge in position to be used as a foot scraper, so that he could scrape off his boots as he entered the kayak.

When a kayaker starts paddling with a Greenland paddle, the paddle tends to flutter during the first few strokes. This fluttering, or side-to-side motion, of the paddle results from the phenomenon known as vortex shedding, caused by alternating loss of the vortex created along each edge as the paddle is drawn through the water. As long as there are equal vortices along each edge, the paddle will draw straight back, but if the blade angle changes enough for one vortex to drop off, or shed, the unequal pressure kicks the blade to one side, making it flutter. The vortices can be seen by standing at the edge of a swimming pool or on a low dock or pier and drawing a paddle through the water.

Once the kayak gets to cruising speed, the narrow paddle is almost effortless to use and does not flutter. The kayaker grasps it where the short loom meets the long blades, wrapping the thumb and forefinger around the loom and the other fingers around the blade. Thus the hands grip the paddle about the same distance apart as the hands normally hang at one's sides. The upper arms hang naturally near the kayaker's ribs at the sides, so that the elbows are not lifted out to the side, but pass near the ribs with each stroke. The forearms are held low and more or less parallel to the water surface when paddling. As seen from in front or behind, the angle at which the upper blade is lifted above the water does not exceed more than about 30 degrees with the water. The trunk or torso twists slightly with each stroke. As long as the speed is kept at three or four knots, a Greenland kayak is very easy to paddle for hours at a time. The narrow blade seems to have just the right amount of slip at cruising speed. This makes it easy to maintain a paddling cadence of about 60 strokes a minute, by counting "one thousand and one, one thousand and two" as each left and right stroke is made.

Another noticeable feature of the Greenland paddle is its buoyancy. This can be demonstrated by holding the paddle vertically in the water, alongside the kayak. In 1995, I watched a Greenlander amuse himself in the harbor at Sisimiut by throwing his paddle like a spear close beside his kayak. When he threw the paddle into the water at an angle of 45 degrees, it would sink almost out of sight, and then buoyancy would cause it to jump back into his hand.

The Greenland paddle is very effective for ordinary paddling and is unsurpassed in performing capsizing maneuvers. The long narrow blade functions as a hydrofoil, just as the long narrow wings of an albatross or a sailplane are very effective airfoils. When a Greenland paddle is swept through the water at a slight angle of attack, it generates lift. The same applies to sculling movements, when the paddle is oscillated back and forth to generate lift.

Seal catchers can also use their paddles as hydrophones to attract seals. Seals have keen hearing, so to make the seal curious enough to surface, a seal catcher will hold his paddle across his lap with one end extended into the water. Then he leans forward so that his mouth is near the flat side of his paddle blade and make a sound as if spitting a hair out of his mouth. Ludwig Quist, a veteran seal catcher from Illorsuit, told me that this noise could be heard by a seal at a distance of about 600 feet (200 m).

When paddling against a strong headwind, a Greenlander can shift the paddle lengthwise with every stroke, so that the advancing blade does not create as much wind resistance. This is done by gripping the paddle in the middle with forefingers touching each other as they wrap around the paddle. For a stroke on the left side, the left hand stays fixed on the loom and the right hand slides out the blade. At the end of the stroke, the right hand is brought back to near the middle of the paddle to become the fixed or control hand for a stroke on the right side. Thus the paddle appears to move lengthwise with each stroke toward the side on which the stroke is being made.

The "storm paddle," which is specially made for this method of paddling, has a short loom that is only long enough to accommodate the two hands touching each other. Most Greenlanders prefer a regular paddle length and seldom find it necessary to use the shifting stroke described above.

ORIGINS OF THE GREENLAND TECHNIQUE As the medieval period ended in Europe, there was increasing interest in the exploration of the Davis Strait and the Greenland region. Much of this activity originated in Denmark. The Danish Arctic Expeditions of 1605 and 1606 brought Greenlanders and their kayaks back to Denmark.

These expeditions gave the Danish public an opportunity to see kayakers show their prowess: "they held their own in a race against a boat of sixteen oars" (Gosch 1897:xci).

This was probably the first exhibition of Greenland kayakers giving a demonstration of their capsizing maneuvers outside Greenland: "three of them, in their boats, performed a kind of dance, cutting figures with their kayaks in a wonderful manner" (Gosch 1897:xcii). These impressive performances were probably the origin of European interest in recreational kayaking: "The King, himself a devoted sailor, took much interest in their performances with their boats, and had one built on the Greenland pattern, but arranged for two men" (Gosch 1897:xcii).

King Christian IV of Denmark has not generally been given credit for being the father of recreational kayaking. Instead, that honor was bestowed on a London barrister, John MacGregor, about 260 years later, for describing his voyages in decked canoes in his *Rob Roy* series of books in the 1860s.

It would be reasonable for an observer in 1600 to describe a demonstration of Greenland capsizing maneuvers as "a kind of dance." Europeans of the 17th century would have had no way of knowing that these maneuvers were, in fact, for saving one's life in the dangerous occupation of being a seal catcher. Not until more than a century later did detailed descriptions of Greenland life by 18th-century observers such as Hans Egede (1741) and Otto Fabricius (1962) begin to appear. The Greenland kayak, while perfect for the jobs at hand, is very unstable. Yet it must operate in the open ocean in a very hostile part of the world. If the kayaker cannot recover to an upright position, he or she will drown. If the kayaker leaves the kayak, he or she will die of hypothermia in a few minutes. The capsizing maneuvers are necessary to stay alive.

Ten different capsizing maneuvers were described in considerable detail as early as 1767 by the Moravian missionary David Crantz. Crantz also gave an excellent description of Greenland kayak hunting approximately a century before the use of firearms had significantly influenced hunting technique. His description, taken from *The History of Greenland*, is quoted at length.

§ 8. The little Man's-boat, called in Greenlandish *kaiak*, is 6 yards in length, sharp at head and stern, just like a weaver's shuttle, scarce a foot and half broad in the broadest middle part, and hardly a foot deep. It is built of a keel like a slender pipe-staff, long side-laths, with cross hoops not quite round, bound together with whale-bone, and is covered over with some fresh-dressed seal's leather as the women's boat; only the leather incloses it like a bag on all sides, over the top as well as beneath. Both the sharp ends at head and stern are fortified with an edge of bone, having a knob at top, that they may not receive damage so soon by rubbing against the stones. In the middle of the covering of the Kaiak there is a round hole, with a rim or hoop of wood or bone, the breadth of two fingers. The Greenlander slips into this hole with his feet, and sits down on a board covered with a soft skin; when he is in, the rim reaches only above his hips. He tucks the under-part of his water-pelt or great-coat so tight round this rim or hoop of the kajak, that the water can't penetrate any where. The water-coat is at the same time buttoned close about his face and arms with bone-buttons. On the side of the Kajak, the first described lance lies ready under some straps fastened across the kajak. Before him lies his line rolled up upon a little round raised seat made for it; and behind him is the seal-skin bladder. His *pautik* or oar, (which is made of solid red deal, strengthened with a thin plate three fingers broad at each end, and with inlaid bone at the sides) this he lays hold of with both hands in the middle, and strikes the water on both sides very quick, and so regular as if he was beating time. Thus equipped, away he goes a fishing, and has as high a conceit of himself as any Mr. captain on his ship. And verily one can do no other than survey the Greenlander in this his parade, with admiration and pleasure; and his sable sea-vestment, spotted with many white bone-buttons, gives him a stately appearance. They can row extremely fast in them, and upon occasion, if a letter requires expedition from one colony to another, they can perform 20 or even 24 leagues [about 60 to 72 miles, or 97 to 116 km] a day. They fear no storm in their kajaks. As long as a ship can carry its top-sail, even in stormy weather, they are not frighted at the boisterous billows, because they can swim over them like an arrow, and even if a whole wave breaks over them, yet presently they are again skimming along the surface. So it is with the kaiak. If a wave threatens to overset them, they counteract its force and keep themselves upright on the water by their oar. Nay even if they are overturned, they give themselves such a swing with their oar, while they lie with their head downward under water, that they mount again in their proper posture. But if alas! they lose their oar, they are commonly lost, unless any one is near at hand to help them up (Crantz 1767, vol. I:150–151).

§ 9. Some Europeans have advanced so far, after a great deal of application and labour, as to be able to divert themselves in the Kajak when the weather and water is still, but seldom are they qualified to fish in it, or to help themselves in the least danger. Now as the Greenlanders are endued with an art and dexterity herein quite peculiar to themselves, which a man must admire with a mixture of panic and pleasure; and as they are obliged to provide all their maintenance in these little cock-shells, in which they are exposed to so much danger, that many a one perishes in the deep; therefore I hope it will not be disagreeable to read some of the *manœuvers* and exercises, which the Greenlanders must learn from their youth up, in regard to recovering themselves and the boat after being overturned in the water. I have taken notice of ten methods of practice, though very likely there may be more.

1. The Greenlander lays himself first on one side, then on the other, with his body flat upon the water, (to imitate the case of one who is nearly, but not quite overset) and keeps the balance with his pautik or oar, so that he raises himself again.

2. He overturns himself quite, so that his head hangs perpendicular under water; in this dreadful posture he gives himself a swing with a stroke of his paddle, and raises himself aloft again on which side he will.

   These are the most common cases of misfortune, which frequently occur in storms and high waves; but they still suppose that the Greenlander retains the advantage of his *pautik* in his hand, and is disentangled from the seal-leather strap. But it may easily happen in the seal-fishery, that the man becomes entangled with the string, so that he either cannot rightly use the *pautik*, or that he loses it entirely. Therefore they must be prepared for this casualty. With this view

3. They run one end of the *pautik* under one of the cross-strings of the kajak, (to imitate its being entangled) overset, and scrabble up again by means of the artful motion of the other end of the *pautik*.

4. They hold one end of it in their mouth, and yet move the other end with their hand, so as to rear themselves upright again.

5. They lay the *pautik* behind their neck, and hold it there with both hands, or,

6. Hold it fast behind their back; so overturn, and by stirring it with both their hands behind them, without bringing it before, rise and recover.

7. They lay it across one shoulder, take hold of it with one hand before, and the other behind their back, and thus emerge from the deep.

   These exercises are of service in cases where the *pautik* is entangled with the string; but because they may also quite lose it, in which the greatest danger lies, therefore,

8. Another exercise is, to run the *pautik* through the water under the kajak, hold it fast on both sides with their face lying on the kajak, in this position overturn, and rise again by moving the oar *secundum artem* on the top of the water from beneath. This is of service when they lose the oar during the oversetting, and yet see it swimming over them, to learn to manage it with both hands from below.

9. They let the oar go, turn themselves head down, reach their hand after it, and from the surface pull it down to them, and so rebound up.

10. But if they can't possibly reach it, they take either the hand-board off from the harpoon, or a knife, and try by the force of these, or even splashing the water with the palm of their hand, to swing themselves above water; but this seldom succeeds.

They must also exercise themselves among the sunken rocks; where the billows toss and foam excessively, and where they may be driven upon the rocks by a double wave besetting them behind and before, or on both sides, or may be whirled round several times, or quite covered over with the breaking surges. In these cases they must keep themselves upright, by artfully balancing the boat, that they may stand out the most violent storms, and also learn to land safe ashore in the midst of the tempestuous waves.

If they overturn and lose all means of helping themselves, they are wont to creep out of the kajak while under water, put up their head and call to any one that is near to help them. But if no one is within call, they hold by the kajak, or bind themselves to it, that somebody may find their body and bury it.

Every Greenlander is not capable of learning all these ways of oversetting and rising again; nay there are good seal-fishers that cannot rise again in the easiest way. Therefore many men are cast away in the seal-fishery, which I will now describe (Crantz 1767, vol. I:151–153).

§ 10. ... In this exercise [seal hunting] the Greenlander is exposed to the most and greatest danger of his life; which is probably the reason that they call this hunt or fishery *kamavok*, i.e. the Extinction, viz. of life. For if the line should entangle itself, as it easily may in its sudden and violent motion, or if it should catch hold of the kajak, or should wind itself round the oar, or the hand, or even the neck, as it sometimes does in windy weather, or if the seal should turn suddenly to the other side of the boat; it can't be otherwise than that the kajak must be overturned by the string, and drawn down under water. On such desperate occasions the poor Greenlander stands in need of all the arts described in the former Section, to disentangle himself from the string, and to raise himself up from under the water several times successively, for he will continually be overturning till he has quite disengaged himself from the line. Nay when he imagines himself to be out of all danger, and comes too near the dying seal, it may still bite him in the face or hand; and a female seal that has young, instead of flying the field, will sometimes fly at the Greenlander in the most vehement rage, and do him a mischief, or bite a hole in his kajak that he must sink (Crantz 1767, vol. I:153–155).

Crantz, along with Otto Fabricius, was one of the few early observers who understood Greenlandic kayaking technique. He realized that the specialized method of hunting with the harpoon and float developed in Greenland was effective, but dangerous. Other northern kayakers also used the harpoon and float. Among them were the Bering Sea kayakers from Bristol Bay to Bering Strait and the kayakers of eastern Canada. Their hunting season for this activity was shorter than in central West or South Greenland, so they were not exposed to the same danger for as long a time. Nor did they use the throwing stick with their large sealing harpoons, as did the Greenlanders. The throwing stick increased the force of the throw, but it added a complication: at the critical moment; the kayaker had to throw the harpoon, put the throwing stick between his teeth, reach back and grab the float and throw it in a stiff-armed motion free of the kayak. The skilled seal catcher accomplished this with one continuous movement of one arm, while the other arm held the paddle to steady the kayak. If the line or float snagged any part of the kayak, the kayak might be pulled underwater by a large seal.

The Greenlanders developed a tray for their harpoon line that stood on three legs elevated off the deck (Figure 1.8). This freed deck space for the implements on the foredeck and kept the coiled harpoon line from being washed overboard by a wave. Because the Greenland harpoon line tray stood above the deck, it was more likely to get entangled in a harpoon line or otherwise cause a problem. The

harpoon line trays were made to break away if snagged, but they could still create distractions that could result in a capsize. These are just some of a number of factors about seal hunting with the harpoon and float that made the activity far more dangerous than hunting with bladder darts, as were widely used in Alaska.

Greenlanders also had the problem of hunting among icebergs, which can calve suddenly, causing huge waves that could capsize a kayak. They needed to develop more skillful and effective kayaking techniques and more capsize rescue methods than many other northern kayakers.

Most Greenlanders identify about 30 different capsize maneuvers; however, there are variations of most of the maneuvers that might be seen as different maneuvers by different kayakers. Each maneuver can be done either clockwise or counterclockwise but is counted as a single method of bracing or rolling.

According to language expert Per Langgård, the generic Greenlandic term for any capsize maneuver is *kinngusaqattaarneq* (noun) or *kinngusaqattaarpoq* (verb). This includes all sculling paddle braces that involve lying at the side of the half-capsized kayak. Each maneuver has a specific Greenlandic name as well. The maneuvers discussed below are from the Greenland Kayaking Association (Qaannat Kattuffiat) and courtesy of Hans Kleist-Thomassen. Together, these maneuvers are valuable ethnographically and are the basis for judging

Figure 1.8—Kayak with white cloth shooting screen and tray for the harpoon line, Uummannaq, West Greenland, c. 1900. Photographer Alfred Berthelsen, no. Al4067. Courtesy Danish Arctic Institute, Copenhagen.

the Greenland Kayak Championships. These contests were held every other year on odd-numbered years until 1993, when they became annual. In 2000, the championships were opened to outsiders.

TRAINING AND TECHNIQUE

Most of the training of Greenland kayakers revolves around the central theme of capsize prevention and recovery technique. Seawater temperature along the coast of West Greenland is often about 35 degrees Fahrenheit (2°C), even in the summer, so it behooves kayakers to be self-reliant. The extremely short time required for outside assistance to reach a capsized victim in such cold water underscores the need for kayakers to be trained in capsize prevention, as well as in self- and group-rescue methods.

Practice begins even before a trainee gets into a kayak. In 1992, I visited the Qajaq clubhouse at Nuuk (Map 3) and was shown a training device that is used for teaching beginners how to condition their balance reflexes. The device is essentially a side-rocking cradle that is just wide enough and long enough to sit upon with the legs outstretched. It is flat on top and rounded on the bottom, resembling half a cylinder. The flat part is where the trainee sits. The "kayaker" sits on top and, with a stick the length of a kayak paddle held in both hands across the lap, attempts to balance the device. Because the training cradle is much tippier than any Greenland kayak, one soon learns that in order to maintain balance without bracing on the floor with the stick, it is essential to keep one's balance point, or center of gravity, which is located near the navel, directly above the point at which the rounded bottom of the cradle touches the floor. Thus one learns to retrain the balance reflexes, which have grown accustomed to leaning the torso by buttock pressure against a seat. In a narrow kayak without paddle contact on the water, balance must be maintained a different way: by keeping the torso vertical and keeping the center of buoyancy of the kayak under the center of gravity of the kayaker. Every successful canoeist or kayaker learns this lesson—sometimes after capsizing—but the Qajaq club method makes it easy and safe. By the time a person who has practiced on the training cradle gets in a kayak, her or she has already mastered the problem of balancing without a paddle. This helps trainees become more confident and reduces the time required to get used to sitting in a real kayak.

The next phase of training is familiarization with the kayak. This can be done locally or at a kayak training camp at a remote location. The latter has the advantage of focusing attention on training because there are curious onlookers and other distractions along the waterfront near towns. Training at a camp does involve travel to the camp and time away from home and job. I observed training sessions at a lake near Sisimiut in 1985, which were part of the activities at the first national meeting of the Greenland Kayaking Association. On that occasion, the sculling paddle brace and standard Greenland roll were taught to young Greenlanders in a lake, which was warmer than the water in the harbor or Davis Strait. Two methods of teaching were used, depending on the water depth. Some kayakers used the middle of the lake with the instructor in an inflatable boat, and some were trained in shallow water at the shore with the instructor wearing waders. In either case, the instructor could hold the gunwale or the bow of the kayak and assist the kayaker if necessary.

The following year, 1986, marked the first year that formal camps were held for Qajaq club members. The camps were held on a fjord near Nuuk, and veteran seal catchers such as Manasse Mathaeussen, Gaba Bernhardsen, and Karl Samuelsen were there to train students from all over West Greenland, the object being for these students to return to their local clubs and act as instructors to train others. Since that year, many more training camps have been held. In 1990, George Gronseth of Seattle attended a training camp in South Greenland and gave his impressions as a recreational kayaker in a *Sea Kayaker* magazine article (Gronseth 1992:46–55).

In 1992, American Gail Ferris and I attended the training camp at Ikamiut, an abandoned village on Tuno (formerly called Hamborgersund). Ikamiut is located 15 miles north of Maniitsoq, West Greenland (Map 3). There were accommodations for about 40 people in a lodge that had been built on the site. The front porch faced the morning sun and overlooked the training area, so it was a popular gathering place.

Two ropes were suspended across the porch at approximately chin height (Figure 1.9). The ropes touched, but could be separated to form a seat, with the ropes passing beneath the thighs as the person faced one of the ends. The hands grasped the ropes before and behind so that a secure seat was provided with the hips between the ropes and the calves hanging outside. Then the person would "capsize" and spin around several times. Some of the Qajaq club members are adept at this game, which is an ancient Greenland pastime associated with kayak training that has been passed down from generation to generation. There are many maneuvers that can be done on the ropes, and the activity shows student kayakers how body movement can help capsize recovery.

Figure 1.9—Arrangement of equipment for Greenlandic rope gymnastics. One pace = approx. 3 feet (1 m).

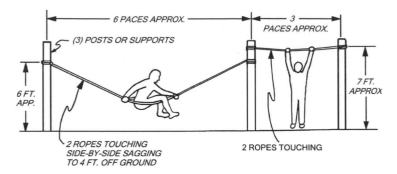

The daily routine at the training camps begins with group exercises that include stretching; flexing the legs, arms, and body; jogging; and so on. The flexing exercises prepare students for the flexibility that will be needed for capsize prevention and recovery when seated in a kayak.

Students are taught how to get in and out of a Greenland kayak. To enter the kayak, the kayak is placed in the water parallel with the water's edge and the paddle is placed across the foredeck in front of the cockpit. The paddle is usually extended on the water side of the kayak, either beneath the foredeck straps or with the hand that is to be on that side clutching both paddle and deck straps to provide a steadying outrigger. Then, carefully wiping the feet so as not to track mud or pebbles, the kayaker, leaning forward and holding the foredeck straps and paddle with the water-side hand while placing the other hand on the shore, steps into the center of the kayak. When both feet are in, the body is lowered to sit on the afterdeck while continuing the hand support. Next, the hips are lowered into the cockpit, which usually requires some wriggling and pulling on the foredeck straps while steadying the kayak. Once inside, the kayaker can secure the spray apron to the cockpit rim. As the student progresses in his or her training, a waterproof jacket will be worn for maneuvers that involve capsizing.

With the instructor standing beside the kayak in water that is hip or waist deep, the first part of practice is devoted to familiarization, stability, and use of the paddle in different positions. The instructor has the student learn balance by keeping the paddle out of the water held horizontally in the hands while the instructor rocks the kayak up and down and from side to side by holding the bow. The student soon learns the importance of keeping the torso vertical and control-

ling roll with the hips and legs. As instruction progresses, the student learns to lean forward and backward and to turn the torso while still holding the paddle out of the water.

This is followed by basic instruction on the use of the paddle for simple strokes and bracing maneuvers. The student learns that the same forces that propel the kayak—that is, pushing on the upper hand and pulling on the lower hand with one blade in the water—will also rotate the kayak around its longitudinal axis. The student is then taught the fundamentals of gaining lift on the working blade by sculling and sweeping the blade across the water surface to gain lift, keeping the leading edge of the blade slightly higher than the trailing edge. Various blade angles are tried in order to demonstrate how the blade stalls and loses lift when it is at too great an angle of attack. The student also learns how fast and how far to move the blade during the sculling stroke.

After the instructor is certain that the student is ready for the next step, they will proceed to the *innaqatsineq*, or sculling on the back paddle brace, which requires that the student wear a waterproof kayak jacket that is fastened about the face, wrists, and cockpit rim to keep out water.

CAPSIZE MANEUVERS Most of the maneuvers discussed below are performed to both the left and right sides, and so require both stamina and skill. These maneuvers testify to the Greenlanders' successful effort to keep their technique traditions alive through competitive sport.

Although the ability to roll a kayak has not been documented in Canada, it was probably known there at one time. I base this opinion on the migrational route of the Thule Inuit, documented archaeologically (McGhee 1984). The Thule culture originated in northern Alaska and spread eastward across arctic Canada to Greenland; it is known that Thule people had kayaks (McGhee 1984:369, 371); however, their kayaking techniques are unknown. There is no doubt that the ability to right a capsized kayak by means of the paddle was widely known in Alaska from Prince William Sound (Birket-Smith 1953) to Point Barrow.

In studying kayaking technique among the last generation of Alaskan and Greenland Inuit who used the kayak extensively for seal catching (i.e., before 1950), I noticed a surprising similarity in the basic methods of rolling a kayak in northern Alaska and in Greenland. Although Alaskans used double-bladed paddles for some of their paddling activities, the Alaskans who were studied always used a single-bladed paddle for rolling. When Greenlanders extend the double-bladed paddle for rolling, it becomes, in effect, a single-bladed paddle. Only the extended blade is used as a working blade. The other end serves as a lever.

The similarity of the starting position for a clockwise roll as done in Kotzebue Sound, Alaska, as compared with a Greenland method is as follows. The Kotzebue Sound kayaker holds the paddle pointing downward on the side toward the capsize. The Greenlander holds the paddle pointing downward and forward on the side toward the capsize. In both cases, the paddler leans toward the paddle to capsize. This brings the kayak to an upside-down position, with the paddle sticking up out of the water. This position has an advantage over starting the next part of the recovery with the blade in the water, because the paddle can be moved more quickly through the air. By hitting the water at a higher rate of speed, the paddle develops more lift for the recovery.

The Kotzebue Sound rolling method resembles the Greenland method much more than it resembles the standard method used at King Island, Alaska, even though King Island is only about 150 miles (240 km) southwest of Kotzebue Sound. Greenland, of course, is roughly 3,000 miles (4,800 km) to the east. I suggest that both the Kotzebue and Greenland methods stem from a common ancestor with a Western Thule source. If this hypothesis is correct, then at least one rolling method was common all the way from Alaska to Greenland, but eventually fell into disuse in Canada. Ulrik Pivat, a seal catcher from Kuummiut, East Greenland, described the Greenland standard method cited above as having been used in both West Greenland and the Ammassalik district of East Greenland (Map 2). Therefore, I am confident that it is not a new or imported innovation.

In the absence of hard evidence, any discussion of similar Western and Eastern rolling methods is speculative. The capsize maneuvers described below are the methods being taught by members of the Greenland Kayaking Association at their training camps and classes, which have been held at various places in Greenland. The following photos, taken by Vernon Doucette, are from a 1994 demonstration for the military at Fort Devens, Massachusetts, by two Greenlanders, Ove Hansen and Pavia Tobiassen.

Sculling on Back (*Innaqatsineq*). The standard sculling paddle brace (Figures 1.10, 1.11) is called *innaqatsineq*, which means "to lie a little at the side." The kayaker extends the paddle in the direction

of the capsize and sculls for support while lying on the back, keeping the torso and head mostly immersed with the face above water. By sculling back and forth with the paddle, the kayaker can regulate the amount of support required and lie on the surface for several minutes if necessary. This maneuver is usually the first one done as a warm-up when giving a kayak rolling demonstration. The *innaqatsineq* is one of the most valuable maneuvers in the Greenlander's repertoire. It is probably the first maneuver described by David Crantz in *History of Greenland* (Crantz 1767, vol. I:152, item 1), although it is possible that Crantz was referring to another sculling paddle brace called *palluussineq* (sculling on the chest), which will be described below.

A common use for any sculling paddle brace is to arrest a capsize before one has turned completely upside down. It is much easier to arrest and recover from a partial capsize if the kayaker can initiate action before his or her head goes underwater. Another application for various sculling maneuvers is in beam seas (waves from the side), when there is danger of being capsized by a large wave. The kayaker can lean into the face of a big wave to prevent a capsize and turn the bottom toward the trough. Wave action then tends to right the kayak instead of rolling it down the wave face.

In order to teach *innaqatsineq*, an instructor selects water with a gradually sloping bottom that is from hip to waist deep. Wearing waders, the instructor stands in the water by the kayaker on the side opposite the direction of the capsize and controls the kayak by holding the upper gunwale. Thus the instructor is out of the way,

Figure 1.10—Sculling on back (*innaqatsineq*).

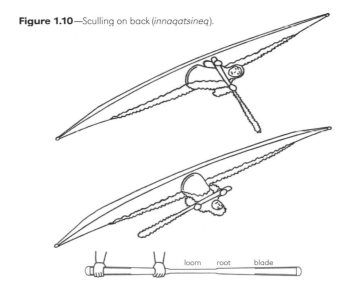

loom   root   blade

Figure 1.11—Ove Hansen demonstrates sculling on back (*innaqatsineq*) to the right. Courtesy Vernon Doucette, 1994.

yet can keep the kayak from falling on the kayaker by holding the upper gunwale as instruction proceeds. The instructor explains how to control the kayak by twisting the pelvis away from the direction of capsize and keeping an upward pressure by bringing the low knee (i.e., the right knee if one is bracing to the right) toward the face. This keeps the kayak leaning slightly away from the kayaker when the craft is on its beam ends (see glossary).

The instructor also explains how to capsize to initiate the maneuver. One method of capsizing is to extend the paddle toward the capsize and begin sculling on the water surface to gain lift while twisting the torso and falling outward and backward. Another method is to extend the paddle toward the capsize at chest level, holding it blade-on-edge across the chest with the elbows bent to bring both hands near the shoulders, palms facing forward against the flat of the inboard blade; then the kayaker twists the torso at the waist *away* from the direction of capsize about 45 degrees from the longitudinal axis of the kayak, as seen from above. The kayaker now faces 45 degrees from the bow, and will capsize straight backwards, which will bring the kayaker 45 degrees from the opposite side afterdeck. As the backward capsize commences, the extended paddle blade slaps the water, which slows the capsizing motion, then the kayaker immediately begins sculling to maintain support. The kayaker, as seen from above, is now lying on his or her back with the torso immersed and the chest and face at the water surface. The torso makes an angle of about 45 degrees with the kayak, which is just short of being on its beam ends, leaning away from the kayaker. The paddle has remained in the same position relative to the torso so it extends outboard and forward on the surface at an angle of about 45 degrees as seen from above.

The sculling motion is initiated as soon as the paddle hits the water. Using the inboard hand—which is holding the inboard end of the paddle—as a pivot point, the kayaker moves the paddle back and forth like an oscillating hydrofoil through a short arc across the surface. The paddle blade's leading edge is inclined upward at an angle of 5 to 10 degrees to gain lift. This angle of attack is reversed at the end of each stroke by a flick of the wrists. The *apparent path* of the stroke is back and forth, but the *real path*—as seen from the working end of the paddle—resembles an elongated figure eight laid on its side. This is because when weight is put on the paddle, it sinks at the end of each stroke and regains position on the return stroke. The angle of attack that the blade makes with the water is relative to the path of the blade, not the water surface. Too steep an angle of attack can cause the blade to stall and lose lift.

Kayakers must avoid an instinctive need to raise the head above the water surface, and instead to keep as much of the body as possible immersed because the water supports the kayaker to a great extent and helps to make the sculling almost effortless. With practice, a skillful kayaker only has to scull an almost imperceptible amount. Veteran kayaker Manasse Mathaeussen told me that in calm water he could take a nap while doing this maneuver.

Some kayakers do this maneuver without as much twist in the torso as described above, so that their side is in the water instead of their back. Others twist their torsos more, so that the shoulders are parallel with the longitudinal axis of the kayak. This brings the paddle forward until it is also parallel with the axis of the kayak, and the torso makes a 90-degree angle with the kayak as seen from above. By experimenting with various positions, the kayaker learns that as the torso is moved outward from the kayak, it becomes increasingly important to keep as much of the body as possible immersed. This particularly applies to the head, because there is a natural tendency to get the face as far out of the water as possible. But by learning to look straight up, the kayaker can be comfortable while keeping only the nose, mouth, and eyes above the surface. The hands have to be at, or just above, the surface, and much of the abdominal area has to be out of the water because it cannot be avoided. Since the abdomen is relatively close to the axis of rotation, it is not as critical to keep it in the water as it is for the chest or the head, which are farther out.

During the initial practice of this maneuver, the instructor holds the upper gunwale of the kayak from the far side. When the instructor is satisfied that the student has mastered it, the hands are raised to show the student that no assistance is needed, which gives the student confidence and pride.

To recover from the sculling position, the paddle is swept aft and as it approaches a point 90 degrees from the longitudinal axis of the kayak, the kayaker brings the lower knee toward the head to rotate the kayak upright. The opposing forces on the hands are increased, levering the kayaker up, but care is taken to rotate only the kayak so that the head does not leave the surface until the kayak is almost upright. This avoids having to lift kayak and kayaker at the same time. The torso is leaned toward the afterdeck as it is lifted from the water surface, which keeps the weight low and near the lengthwise axis of the kayak. As the roll upward is completed, the kayaker returns to an erect sitting position. Throughout the maneuver the hands have been near the shoulders, palms facing upward to grip the inboard paddle blade. In lowering the paddle, the forearms are moved to bring the hands forward and downward to the deck, still gripping the paddle in the same manner. The paddle is now against the deck with the palms facing downward holding the inboard blade, while the extended outboard blade serves as an outrigger to steady the kayak. The position of the torso on the surface during this maneuver can vary about 90 degrees relative to the axis of the kayak—that is, from sculling on the side while leaning on the afterdeck to sculling on the back with the torso 90 degrees from the axis. The two extreme positions might be perceived as *side* and *back* sculling paddle braces, but the Greenlanders seem to perceive them as variations of the same maneuver, *innaqatsineq* (sculling on the back). They usually hold the torso about halfway between the extreme positions, or just far enough out from the kayak to get the shoulders in the water to gain buoyancy with the inboard elbow against the afterdeck or the kayaker's side.

Sculling on Chest (*Palluussineq*). Although sculling while lying with one's back on the surface is useful in many situations, there are times when a seal catcher might get a harpoon line caught on the paddle or kayak. The kayaker could be pulled under while facing the direction of the capsize (facedown). So the Greenlanders developed a maneuver called *palluussineq*, which is done by sculling on the chest while facing the direction of the capsize (Figure 1.12). This maneuver is usually the second one done as a warm-up when giving a kayak rolling demonstration, immediately following the first warm-up maneuver *innaqatsineq* (sculling on the back), described above. The *palluussineq* is of great importance to Greenland seal catchers, who hunt from the kayak with a harpoon and float. The seal, which nowadays is almost always shot before being harpooned, is approached with the bow turned slightly away from the side on which the harpoon is thrown. The seal dives as soon as it is hit by the harpoon. As the harpoon line reels out of the line tray, it might catch the kayak, a piece of deck equipment, the paddle, or the kayaker and capsize the kayak in the direction of the quarry—that is, the direction the kayaker is facing. The same thing might happen if the float gets caught and cannot be thrown out quickly. By sculling on the chest, a kayaker may be able to keep from being pulled under. Sculling on the chest is also useful in preventing any capsize on the side toward which one is facing.

This maneuver is executed by twisting the shoulders parallel to the longitudinal axis of the kayak, so that the kayaker faces the direction of the capsize. This brings the elbow that was on the side of the capsize over the after edge of the cockpit hoop; the other elbow is now over the forward edge of the hoop. The paddle is extended in the direction of the capsize, straight out to the side and on, or just above, the water surface. The forward hand grips the end of the paddle that is closest to the boat from above with the palm facing the flat side of the blade. The aft hand grips the same blade toward the middle of the paddle near the root of the blade, also from above with the palm facing the blade. The thumbs are on opposite sides of the blade. The kayaker leans toward the paddle and capsizes, putting weight on the aft hand, which is now outboard. The inboard, or forward, hand becomes the pivot point for the sculling motion. The arms are held straight and rigid, with upward pressure on the end of the paddle nearest the boat and downward pressure on the outboard hand. Veteran seal catcher Karl Samuelsen prefers to grip the paddle with the outward elbow bent. This is a more natural position in an emergency.

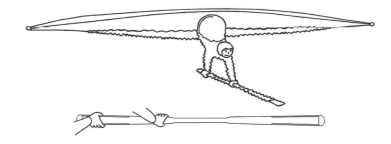

Figure 1.12—Sculling on chest (*palluussineq*).

The sculling motion is started at the beginning and kept up as weight is put on the outboard hand. The inboard end of the paddle sinks as the capsize commences. As the kayak approaches 90 degrees from vertical, the outboard blade tip is just above the surface and the inboard end is an arm's length deep. The kayaker keeps the head bent back so that the chin is at the surface as he or she looks outward. Because of the contorted position, this maneuver is difficult to maintain for more than a few seconds, and the uncomfortable position is made worse because most of the kayaker's head is above water, making it harder to support. Despite these drawbacks, this maneuver is important for a seal catcher to learn in order to survive. To recover from the sculling position, the kayaker increases the opposing forces on the hands and rolls the kayak upward with the low knee while dragging the sculling paddle inward as the torso is raised.

Standard Greenland Roll (*Kinnguffik Paarlallugu*). The standard method of rolling a kayak in Greenland is called *kinnguffik paarlallugu* (Figure 1.13), which means to capsize on one side and come up on the other, according to Nuka Møller, a linguist and secretary to the past president of Greenland, Lars Emil Johanson. Greenlanders consider this the most important method of rolling a kayak because it is so efficient and dependable. *Kinnguffik paarlallugu* is the first method of rolling taught to beginners as soon as they master the *innaqatsineq* (sculling on the back), described above. The ability to perform these maneuvers does not guarantee safety, but it reduces the risk that is inherent in kayaking. The confidence of being able to recover from a capsize goes a long way toward encouraging a kayaker to remain seated in the kayak after an accidental capsize, because to get out in frigid water, even if help is nearby, can be fatal. Experience at being capsized and recovering helps kayakers to stay reasonably calm and to try again if an attempt to roll fails. By having the standard roll available as a means of self-rescue, Greenlanders feel confident to practice more difficult methods of rolling. At practice sessions, one sometimes sees kayakers fail once or twice in some difficult maneuver, after which they shift the paddle in their hands and roll up by using the standard roll.

There are several slight variations on the standard Greenland roll. The method described below is the one taught by the Greenland Kayaking Association. The first step in training for capsizing on one side and coming up on the other is to be capsized under safe conditions and remain underwater for a few seconds. Students know that the instructor is there and will bring them up when they signal. This

Figure 1.13a—Standard Greenland roll (*kinnguffik paarlallugu*). The solid lines represent the starting position for a clockwise roll. The paddle is held blade-on-edge along the starboard gunwale, with one end near the right hip and the other end toward the bow. The kayaker leans forward and faces slightly to starboard. His left forearm is against, or near, the foredeck, and his left hand reaches across the starboard gunwale to grasp the paddle near, but short of, the middle. The right hand holds the paddle near the end, about even with the hip. The palms of both hands pass over the paddle, so that the knuckles are outboard. The kayaker takes a deep breath, leans to starboard and capsizes.

Figure 1.13b—A fish-eye view of the standard Greenland roll. To right himself, the kayaker: (1) Flicks his wrists to swing his knuckles toward his face, thus causing the outboard edge of the paddle to assume a slight planing angle with the water surface. The remaining steps constitute one continuous movement, to be done as quickly as possible. (2) With his hips and right hand serving as pivot points, the kayaker sweeps his forward paddle blade and his torso outward in a 90-degree planing arc on the water surface (as shown in positions 1 through 3) while pulling down on his left hand and pushing up on his right, thus lifting himself to the surface. (3) The kayaker completes the roll by flicking his wrists to flatten the blade angle, then sharply increasing his opposing hand pressures, thus raising himself in a chinning attitude as the paddle blade sinks and is drawn inward. The roll is now complete.

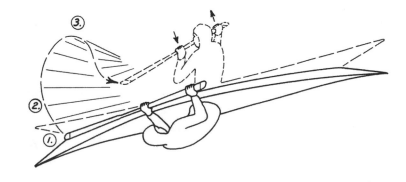

Figure 1.13c—The critical stage of a capsize recovery. The start (solid lines) and finish (dashed lines) of a planing sweep are shown head-on. Success is almost certain if the kayaker has surfaced by the time he has completed the 90-degree sweep. Some minor refinements of rolling are apparent. The left forearm is shown right against the foredeck (a convenient means of orientation), the leading shoulder is nearer the surface (to gain lift when the torso is swung outward), and the hips right the kayak as far as possible while the torso is still partly submerged (to avoid having to lift torso and kayak at the same time).

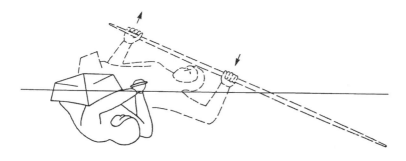

knowledge gives students confidence and helps eliminate panic, which is a major barrier to learning any maneuver in which the head is underwater. The snugness of a Greenland kayak worsens the panic, because students know that if it becomes necessary to get out of the kayak underwater, extra time is needed to wriggle out and reach the surface. The instructor reassures students by gradual exposure to the experience of hanging their heads down in the water. Students are instructed in how to keep a slight pressure on their nostrils and to keep their faces as near the deck of the inverted kayak as possible. Both of these steps help keep water out of the nose. If one hangs underwater with the torso extended downward, there is more water pressure on the nostrils than if the face is nearer the surface. Leaning forward or aft, depending on the maneuver, also keeps the weight and water resistance of the torso near the longitudinal axis of the kayak. This reduces the effort required for the instructor to roll students upright, and it is good practice for getting into the optimum starting or finishing position for rolling maneuvers.

Next, the instructor has students put both arms around the kayak while leaning forward. The palms of both hands are against the bottom of the kayak. By drumming the fingers against the inverted bottom of the craft, students can let the instructor know when to right the kayak.

I have observed several methods that instructors use to capsize and right students in Greenland. The method selected depends upon personal preference, the depth and bottom slope of the water at the practice site, and whatever equipment is at hand. In deep water, the instructor uses an inflatable boat or a similar stable boat and leans over the stern or side to control the kayak by holding the gunwale or the bow. In shallow water, the instructor might wear waders and stand in the water to instruct. (If the bottom slope is such that an instructor can stand in water less than knee-deep to control a kayak by holding it at the bow, the instructor can get by with ordinary boots instead of waders.) A rough rule of thumb is that water under the kayaker must be at least waist-deep in order to perform the standard roll safely. That same depth is adequate for practicing most other rolling maneuvers. Deeper water is required for some of the rolls that require sculling all the way around with the paddle, or if the paddle is extended vertically toward the bottom while sculling.

West Greenland kayaks are deep at the forefoot, which is the point at which the bottom curves upward to meet the bow. This shape enables the instructor to stand at the bow and grasp the bow with one hand and lean over to grasp the forefoot with the other. By moving the hands in opposite directions, the instructor can right a capsized kayak. This is easier if the kayaker leans forward and wraps his or her arms around the kayak. When the student is ready, he or she leans the body in the direction desired to capsize. After a few seconds underwater, the student drums his or her fingers on the bottom of the kayak as a signal to be rolled up. Gradually, the student loses some of the fear of being underwater and gains confidence. From then on it will be possible to concentrate on the manipulation of the body, the paddle, and the kayak in order to recover from a capsize.

The starting position for the *kinnguffik paarlallugu* is for a roll to the kayaker's right, or clockwise. The paddle is held blade-on-edge alongside the starboard gunwale so that it is level with the water surface, with the aftermost end of the paddle near the kayaker's right hip and the other end extended forward near or touching the gunwale. The kayaker leans forward so that his or her face is near the foredeck and turned toward the paddle. The left arm is bent at the elbow and is near or against the foredeck, with the left wrist on the starboard gunwale. The left hand is just outboard of the starboard gunwale, grasping the paddle near the root of the aft blade with the knuckles outboard. The right arm is bent at the elbow, which is kept aft to bring the right hand into position to grasp the aft end of the paddle at the kayaker's right hip, with the knuckles outboard. Thus, the palm of the right hand is against the outboard face of the blade at the widest point. This grip enables the aft hand to control the blade angle in this particular roll.

The kayaker takes a deep breath, leans to the right, and capsizes. The blade-on-edge position of the paddle keeps it from being knocked out of position during the capsize. Gravity does all of the work for the first half of the roll, so the kayaker reaches the fully capsized position simply by maintaining the starting position. After capsizing, the first step of the capsize recovery, or last half of the roll, is to adjust the paddle planing angle. By flipping the wrists back, bringing the knuckles toward the kayaker's face, the paddle is rotated on its lengthwise axis to bring it from the blade-on-edge position to a planing angle, ready for the 90-degree sweep across the surface, which will be the next step. The planing angle is a matter of individual choice and can range from almost flat on the surface up to about 10 degrees. Too much angle will make the blade stall; a negative angle will make the blade dive during the outward sweep. Since the blade tends to sink as weight is put on it during the sweep, the real planing angle is relative to the path of movement, not the water surface, and hence tends to be greater than the kayaker perceives. That is why many kayakers prefer to start with the blade flat on the water surface. The angular velocity is another factor, so some kayakers prefer to get more paddle speed at the start by lifting the hands toward the sky at the start of the sweep, which raises the paddle off the surface. They believe that the surface slap as the blade hits the water also gives some lift.

As the kayaker begins the outward sweep with the forward end of the paddle, the mental picture of the path it will follow is a 90-degree sweep across the surface. The aft or inboard hand is a pivot point and the forward hand, which is sweeping outboard with the paddle, supports the weight as the torso is raised. In the first part of the sweep, the emphasis is on getting enough speed to lift the kayaker's weight. The torso is pivoted at the hips to swing outward with the paddle. As the body swings out, the hips are used to start rotating the kayak up on its beam ends. The trailing knee (the left knee for a roll to the right) is brought up toward the kayaker's face, which aids in rotation. This causes the left thigh to push upward against the deck, which rolls the kayak about its longitudinal axis.

For many years, the dean of Greenland kayakers was Manasse Mathaeussen. He was one of my most valued sources for data from 1985 until his death in 1989. After Manasse's death, his son, Kristoffer, told me something Manasse had told him years before about this roll. Kristoffer said that his father had emphasized the importance of keeping the head down in the water until the kayak is rotated with the lower body so that it falls away from the kayaker. Manasse said that it helps a kayaker to lift the trailing knee and keep the head against the outboard arm during the sweep, especially as the sweep is finished and the body is raised. He said it is also important to keep the inboard elbow against the hip to provide support for the pivot hand.

As the torso and paddle are swung outward to the 90-degree position, the kayak is controlled with the trailing thigh and the hips to keep it rolling ahead of and away from the kayaker. At the end of the sweep, the opposing hand pressure is increased sharply and this brings the kayaker upright, leaning aft and hanging from the paddle in a chinning position. If the roll is not complete at the 90-degree position, it can be saved by extending the sweep through more of an arc. If still more lift is needed, the planing angle can be reversed and the blade swept forward.

The roll is completed with the palms of both hands facing upward near the kayaker's head, so the final movement, as the kayaker brings the body forward to the erect position, is to bring the hands forward and downward to the deck without changing the grip on the paddle. This places the paddle on the foredeck with the palms facing downward and the paddle extended to the left in a bracing position. The maneuver is complete.

A variation on the method described above is to capsize with the paddle held almost as described in the starting position above, except that it is laid flat in the water along the gunwale. The paddle is gripped with the palms facing downward. There is a risk that the paddle will be swept out of position during the capsize, but this position has the advantage of not requiring the kayaker to flip his or her hands to change the blade angle before recovery. The paddle merely has to be swung outward for the 90-degree sweep because the paddle and hands are already in the correct position. Many experienced kayakers prefer this variation and it should be easier for most beginners to learn.

Roll with Paddle Held in Crook of Elbow (*Pakassummillugu*). This roll (Figure 1.14) is done with the paddle held in the crook of the elbow of the arm that sweeps the paddle out from the bow as a hydrofoil to gain lift in a 90-degree sweep aft. Thus, the movement of the paddle is similar to that of other rolls that start at the bow and sweep aft. Before capsizing to the right for a clockwise roll, the kayaker holds the paddle across the kayak in the crook of the left elbow, with the paddle extended to the left. The other, or inboard, end of the paddle is only just outboard of the kayaker's right shoulder, with the right arm bent so that the right hand holds the blade-on-edge paddle near the right shoulder with the palm of the right hand flat against the aft face of the blade.

To get in position for the capsize, the kayaker swings the extended left end of the paddle forward and to the right across the foredeck so that it is blade-on-edge, just outboard and flat against the right gunwale. The paddle is still in the crook of the left elbow with the left forearm wrapped under and outboard the paddle so that the arm points aft toward the kayaker's face. The left hand is on top of the paddle near the kayaker's face. The right hand holds the aft end of the paddle about waist high with the right palm facing outboard against the flat of the blade. The paddle is inclined as seen from the side of the kayak, with the forward end of the paddle held near the right gunwale, blade-on-edge.

The kayaker leans to the right to capsize and holds the above position until fully capsized. The 90-degree sweep and recovery are essentially the same as in the standard roll, except the way the paddle is being held crooked in the elbow of the sweeping arm, which is the left arm for a clockwise roll.

This roll would be useful for a Greenland seal catcher to know if one arm were to become tangled in a harpoon line, yet still had some movement. An example might be the situation of a harpoon line wrapped around one wrist and the neck, in such a way that the wrist could not be moved far from the neck. The *pakassummillugu* maneuver is usually practiced with another maneuver *(paatip kallua tuermillugu illuinnarmik)* that is also meant for a situation in which an arm might be entangled or otherwise unavailable for use in capsize recovery.

Roll with End of Paddle in Armpit (*Paatip Kallua Tuermillugu Illuinnarmik*). This maneuver is practiced with the roll described above *(pakassummillugu)*, in which the paddle is held in the crook of an elbow (Figure 1.15). This roll is quite different in that it is practiced with one arm holding the paddle, with the inboard end of the paddle held in the armpit of the same arm. The arm is held on top of the paddle, which is held flat or parallel with the water surface. With the end of the paddle held in the right armpit, flat side against the underarm, the hand reaching as far as possible down the paddle with the palm down and grasping the paddle, a right-handed kayaker would swing the extended arm and paddle across the foredeck as far forward as possible so that the forward end of the paddle crossed the left gunwale near the bow. Then the kayaker capsizes to the left. All other rolls described above have been to the right, or clockwise,

Figure 1.14a—Ove Hansen demonstrates rolling with the paddle in the crook of the elbow *(pakassummillugu)*. All photos courtesy Vernon Doucette, 1994.

Figure 1.14b—Surfacing with the paddle still in the crook of the arm.

Figure 1.14c—Hansen completes *pakassummillugu*.

Figure 1.15a—Roll with the end of the paddle in the armpit (*paatip kallua tuermillugu illuinnarmik*); photos show a left-handed roll. All photos courtesy Vernon Doucette, 1994.

Figure 1.15b—Capsizing.

Figure 1.15c—Surfacing with free arm counterbalancing.

Figure 1.15d—Completing the roll.

which is the direction seen most often in Greenland, but this roll would preferably be to the left for a right-handed kayaker. After having reached as far forward as possible, the kayaker is leaning well forward during the capsize. As soon as the capsize is completed, the kayaker sweeps the paddle outward and swings the torso outward and backward to finish the roll leaning well back on the afterdeck, paddle extended to the right. The free arm, which is the left arm for a right-handed kayaker, is thrown over the left side and bottom of the kayak as a counterbalance when the kayaker surfaces.

Reverse Sweep with Paddle across Chest (*Kingumut Naatillugu*). When running before the wind in the big swells that can occur in Davis Strait, Greenland kayakers sometimes surf down the face of a big swell with the paddle held aft at an angle to act as a rudder. The kayaker faces the paddle, twisting the torso enough to hold the paddle comfortably in both hands with the after blade piercing the surface well aft of the cockpit. Although the twisted body position tends to be unstable, the vertically held paddle provides support for bracing. Stop for a moment and visualize the paddle and kayak from the bow or stern. Then you can see that the paddle, even though slanted aft, is in a vertical plane when viewed from the bow or stern. A slight twisting of the paddle and/or opposing side pressure with the hands is usually enough to correct tipping before it develops into a capsize.

Once in a while a capsize occurs, and the roll described below (Figures 1.16, 1.17) was developed largely for capsizing while using the paddle as a rudder. This is a popular method of rolling in West Greenland, and it is often performed in practice sessions.

For a clockwise roll, the kayaker holds the paddle in front of the chest with the paddle extended to the left. Both elbows are bent to bring both hands in front of the shoulders, palms facing the flat side of the inboard blade, which is held on edge, still extended to the left.

The next move must be done quickly because the kayaker will be in a very unstable position once it commences and might capsize in the wrong direction. The kayaker turns quickly to the left, swinging the extended paddle across the afterdeck, and hooks it over the starboard gunwale near the stern. At the same time, the head is flung back over the starboard gunwale causing the kayaker, who is now facing left, to capsize backwards. (Because the kayaker cannot see behind during the capsize, her or she must always look aft before doing this roll.) Once capsized, the kayaker sweeps the paddle downward, holding it so that it will develop hydrofoil lift as it

Figure 1.16—Reverse sweep with paddle across chest (*kingumut naatillugu*). The solid lines represent the starting position for a clockwise roll (1). Holding the extended paddle as shown, the kayaker twists the torso leftward to hook the blade over the edge of the starboard gunwale near the stern. The head is thrown back and the left knee is lifted to capsize backward (2). The dashed lines represent the fully capsized position. The paddle is swept forward to complete the roll.

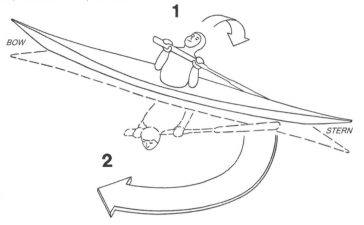

is swept outward and forward, bringing the kayaker up and bracing downward on the paddle.

To Lie Down (*Nalaasaarneq*). Skilled kayakers have a special maneuver that can be used for resting at sea when they become tired and it is impractical to go ashore. According to Greenlander Lone Bech, an avid kayaker and past treasurer of the Greenland Kayaking Association, this maneuver is called *nalaasaarneq*, which means "to lie down" (Figure 1.18). *Nalaasaarneq* enables a kayaker to fall backwards upon the water surface and lie there for several minutes without any movement of the kayak, kayaker, or paddle. The kayak remains heeled on edge so that the kayaker's weight is shifted, which helps relieve the tired muscles and aching joints that can result from sitting several hours in a kayak. Per Langgård and Nuka Møller told me that Greenlanders call the kayaker performing this maneuver *qasuersaartoq*, "the one resting."

Although *nalaasaarneq* is a paddle brace inasmuch as the paddle is used to support the kayaker's position on the surface, it differs from all other paddle braces. In the *nalaasaarneq* maneuver, instead of using an extended paddle to brace against water for support, the paddle is used to lock kayak, kayaker, and paddle into a floating unit that remains stable and without movement. The natural forces acting upon the floating unit are in equilibrium. As long as the water is relatively calm, a skilled kayaker may even take a nap in this

Figure 1.17a—Reverse sweep with paddle across chest. Note that the paddle is hooked behind the starboard gunwale. All photos courtesy Vernon Doucette, 1994.

Figure 1.17b—Ove Hansen starts the forward sweep for the recovery.

Figure 1.17c—Halfway up on the recovery.

Figure 1.17d—Hansen finishes the roll, *kingumut naatillugu,* in a low brace.

Figure 1.18a—To lie down (*nalaasaarneq*). The position that the kayaker assumes as he capsizes backward will be held throughout the maneuver.

Figure 1.18b—View from astern showing the torso twisted as the kayaker lies flat on the surface of the water with the kayak heeled away. Arrows indicate pressure, not movement. Paddle will be shifted outboard to roll up.

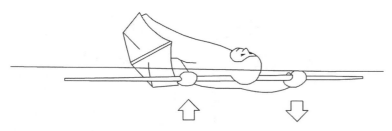

position. One such kayaker was the late Manasse Mathaeussen, who told Qajaq members that when he took a long trip with another kayaker, he insisted that his paddling companion must also know how to perform this maneuver. Mathaeussen taught this maneuver to the Qajaq members in the 1980s. He used to demonstrate *nalaasaarneq* in public performances and was the undisputed master of this maneuver until he retired in 1986, at age 71. Many members of Qajaq can perform *nalaasaarneq* today.

To perform *nalaasaarneq*, the kayaker holds the paddle across the shoulders behind the neck, with the hands outboard of the shoulders on each side, and shifts the hand that will be outboard toward the end of the paddle. Then the kayaker twists the torso so the back faces the direction of the capsize. At the same time, the kayaker leans aft and hooks the paddle outboard the afterdeck gunwale on the side toward the capsize, then quickly throws his or her head back to capsize, while keeping the kayak heeled with the legs and pelvis so that it is inclined away from the direction of the capsize. When the kayaker hits the water, the head goes under, then quickly bobs up. The kayaker, who has remained in the position assumed at the beginning, now floats on the surface, facing upward. To recover from this position, the paddle is shifted lengthwise out from under the kayak and used as in completing other rolling or bracing maneuvers.

As is the case for all capsizing maneuvers, to perform *nalaasaarneq*, the kayaker wears the *tuilik*, a full waterproof kayak jacket, which is sealed around the face, wrists, and cockpit rim (Figure 1.19). A sweater is usually worn underneath the *tuilik* for warmth, but the bulk of the sweater also provides buoyancy. Buoyancy in this and other capsizing maneuvers is also somewhat affected by the salinity of the seawater, which in turn affects stability. I mention this at this point because this slight additional buoyancy may be needed to get this maneuver to work. The shape of the Greenland kayak design is also very important. This will not work with some, perhaps most, recreational kayak designs.

After a kayaker learns how to do this maneuver by keeping paddle pressure against the immersed side of the kayak to maintain the angle of heel, the next step is to practice using the hand to grasp the immersed chine. By grasping the immersed chine and pulling it toward the kayaker, the kayak is rotated on its axis so that it tends to fall away from the kayaker. This accomplishes the same result as applying an upward pressure against the side of the kayak with the paddle. With practice, one can lie beside the kayak without using the paddle. *Nalaasaarneq* is an ingenious example of how buoyancy and stability affect a floating object. With this in mind, I entitled an article in *Sea Kayaker* about this maneuver "The Balance Brace" (Heath 1992). But if Greenlanders were to discuss *nalaasaarneq* in English, they would probably describe it as a "resting" or "reclining" maneuver rather than as a brace. No matter how it is described, *nalaasaarneq* is an impressive maneuver to witness.

Figure 1.19—Ove Hansen tightens the wrist-sealing strap of his *tuilik*. Photo courtesy Vernon Doucette, 1994.

Certain regions along the coast of Greenland are noted for violent storms. Even the most peaceful fjords are not immune from gale-force winds that spill over the mountains surrounding the inland ice cap. Nuka Møller related that an expert kayaker lost his life in a sudden storm in Nuup Kangerlua, near Nuuk airport, in the early 1990s. Sadly, several capable kayakers have been lost in storms in recent years. Sometimes it is impossible to make headway against high winds. A kayaker might be blown offshore by gale winds and then be unable to get back to shore until the storm abates, which could be hours, or even days, later. Such an ordeal can test the stamina, as well as the skill, of the best kayaker. An example of such an incident was recorded on the first British Air Route Expedition:

> In 1931 when at Cape Farewell, H. G. Watkins met an Eskimo who had just returned from being blown to sea for four days in a phenomenal storm. He was much exhausted, because he had to roll constantly due to capsizes from waves which were curling so high that they threw him over endways (Anon. 1934:62).

A shorter description of this incident is in the book *Watkins' Last Expedition* (Chapman 1934:115). To be blown to sea in a storm for four days and survive is a most extraordinary adventure for any kayaker. Yet being lost overnight in storm or fog is always a possibility. Johannes Rosing wrote a dramatic description of a kayaker's fear and confusion in spending New Year's Eve, 1899, lost in heavy seas and blinding snow (see Ataralaa's Narrative, this volume).

Since the danger of being caught offshore in a sudden storm is unavoidable, the Greenlanders developed techniques that enable kayakers to survive by avoiding the impact of the biggest curling waves and by conserving energy in order to buy time. The standard method of rolling is a convenient and reliable way to recover from a capsize, but in stormy seas, a kayaker who rolls up by the standard method can face a secondary capsize. However, one who is caught in a storm and is unable to make headway might still survive until conditions improve by conserving energy with the three maneuvers described below.

Combination Roll and Brace (*Kinnguffik Paarlallugu / Innaqatsineq*). The first of these maneuvers is a combination of the standard Greenland roll, *kinnguffik paarlallugu*, and the standard sculling brace, *innaqatsineq* (sculling on the back), described above (Figures 1.10, 1.13). One who can scull while lying on the back and who can also perform the standard roll can combine these maneuvers and pause on the surface before rolling up. These maneuvers are both efficient and relatively easy to learn. Once they are mastered, starting with the sculling paddle brace and then advancing to the standard roll, they can be practiced in combination to simulate the recovery used in storm conditions.

To practice this maneuver, the kayaker begins by capsizing in the position assumed for the standard roll, *kinnguffik paarlallugu*. After capsizing, the paddle and torso are swept out as in the standard roll recovery. This brings the kayaker to the surface, looking straight up. But instead of completing the recovery, the kayaker lies on the surface and sculls to support the face above water, just as in the standard paddle brace, *innaqatsineq*.

In this maneuver, the kayak is on edge and kept tilted away from the kayaker by keeping the lower, or trailing, knee bent toward the kayaker's face. But in stormy conditions, the kayaker intentionally allows the kayak to fall toward him or her when big seas threaten, so that the bottom of the kayak can take the impact of the curling waves. The kayaker also has the option of remaining on the surface for several seconds or even minutes at a time, then rolling up when ready and immediately leaning forward with the paddle extended in a low brace in order to rest and avoid being capsized again. Although this technique is one of the basic maneuvers of storm survival, it behooves kayakers to prepare for as many situations as possible. Two other methods of recovery are described below.

Storm Recovery Combination (*Kingumut Naatillugu / Palluussineq*). The *kingumut naatillugu* (reverse sweep with paddle across chest) is the second most popular method of rolling. Instead of the paddle being swept aft as in the standard roll, the paddle is swept forward from the afterdeck gunwale. Instead of recovering in a high or hanging brace position as in the standard roll, the *kingumut naatillugu* recovery (Figures 1.16, 1.17) is made in a low brace position—that is, with the palms facing downward. Thus, when used in stormy conditions, the *kingumut naatillugu* works well in combination with the *palluussineq*, or on-the-chest method of sculling (Figure 1.12). In both maneuvers, the torso is over the paddle with the palms facing away from the kayaker, so that the recovery is made in a low bracing position. This makes the two maneuvers compatible when used in combination.

To practice this combination, the kayaker capsizes backwards as shown in the description of the *kingumut naatillugu*. To recover, the paddle is swept forward. When the head breaks the surface with the kayaker facing outward, the kayaker pauses with the kayak on edge and the maneuver becomes the *palluusineq*, or on-the-chest method of sculling (Figure 1.12). As mentioned before, this sculling brace is more difficult than sculling on the back because it is necessary to raise the head and some of the torso out of the water so one can breathe. But it is useful if one capsizes with the paddle extended aft. Once the head breaks the surface, the kayaker has the option of evaluating sea conditions and completing the roll or ducking under after getting a breath and then repositioning the paddle for a different method of recovery. The kayaker might then decide to do the standard roll–standard brace combination. Or, if the kayaker does not wish to pause on the surface and scull, the Greenland storm roll might be selected.

Greenland Storm Roll. This roll (Figure 1.20) and the standard Greenland roll were described in the *Polar Record* in an article titled "The Eskimo Kayak" (Anon. 1934). The article was based largely on data gathered by the Watkins expeditions of 1930–31 and 1932–33. The returning members of the second expedition visited Cambridge and one of them, John Rymill, along with kayak enthusiast J. I. Moore, furnished the editor of the *Polar Record* with data upon which the "Eskimo Kayak" article was based. This description is for the roll to the left, or counterclockwise. All of the other maneuvers described herein are to the right, unless noted. It is only necessary to substitute "right" where "left" appears in the description—and vice versa—in order to describe a roll to the right:

> When the kayak is upside down, the body is bent as far forward as possible until the head is a few inches from the deck of the kayak, and this position is held until the kayak is upright again. The paddle is held near the end in the left hand, and at the centre with the right hand, the paddle lying flat against the side of the kayak and parallel to the surface of the water. The left hand is then pushed as far forward and upwards as possible, while keeping the back of the paddle flat against the kayak until the fingers of the left hand rub against the bottom of the kayak. While the left hand is being pushed forward, the right hand is brought back over the head. This may all be described as the first movement. The second movement is pulling the paddle sharply to the right with the right hand, and raising the left hand until it is

level with the deck of the kayak. The kayaker should then be in an upright position with the body bent well forward (Anon. 1934:61).

The storm roll was used by F. Spencer Chapman, a member of both Watkins expeditions, when he capsized in rough seas on the second expedition: "I came up again at once, without much difficulty, using the storm roll which brings one up in a stable position" (Chapman 1934:283).

The stable position in which a kayaker is brought up by the storm roll results from leaning forward with the head as near the deck as possible and also by having the working paddle blade extended in a low brace at the end of the roll, with the hands gripping the paddle near each side of the deck. Keeping low helps prevent another capsize, since a kayaker who is sitting erect is more likely to be hit squarely by a breaking wave. If one is hit from the side by a wave, the kayak tends to roll. If the wave breaks against the chest, it can damage the kayak as Chapman mentions in the following quote:

> The natives say there is no danger of the kayak being damaged unless the waves actually break against the chest of the man. In parts of the extreme south of Greenland, where they are more used to kayaking in rough seas, when they see an exceptionally dangerous wave coming they capsize on purpose, take the wave on the bottom of the kayak instead of letting it hit them in the chest, and then when the wave is safely past they come up again with the paddle (Chapman 1934:283).

Chapman made another observation about rolling a kayak that can be confirmed by those who have rolled in high winds: "It is very much easier to roll if you go over with the wind and let it help you up

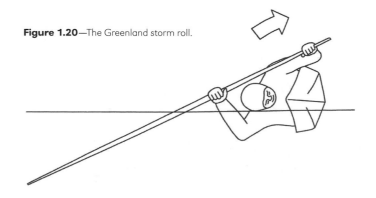

Figure 1.20—The Greenland storm roll.

again. Much less effort is needed if you start to come up again before the kayak has come to a standstill on its way over" (1934:278).

In leaning over to scull, it is better to lean into the wind and come up with the wind. This usually means that if the waves are large, one will be leaning into the slope of the wave instead of the trough; it is much like walking around a steep hill because there is only thin air on the downhill side. But the sea is constantly changing and only careful training can prepare a kayaker for coping with a storm at sea. And no matter how much training and skill a kayaker might acquire, there is always the possibility of getting caught in a storm that can overwhelm any kayaker.

SCULLING ROLLS

Sculling Roll with Paddle under Kayak (*Qaannaq Ataatigut Ipilaarlugu*). Greenlanders have developed several methods of rolling a kayak through a complete revolution by using the sculling motion of their extended paddle. In these maneuvers, the paddle is moved back and forth through a short arc, always keeping the leading edge at a slight angle of attack, in order to obtain lift. The manipulation of the paddle is similar to that used in the sculling paddle braces previously described in the maneuvers called *innaqatsineq* (sculling on the back) and *palluussineq* (sculling on the chest).

The maneuver described below is the most popular of the sculling rolls. According to Greenlander Harald Sandborg, the maneuver is called *qaannaq ataatigut ipilaarlugu*, which means "to use the paddle under the kayak." To practice this roll, the kayaker leans close to the foredeck and passes the paddle under the kayak with the working blade extended horizontally outward on the side away from the direction of the capsize. Thus, for a clockwise roll, the paddle would be passed through the water under the kayak, with the working blade extended to the kayaker's left. The kayaker holds the paddle firmly at both sides of the kayak, with the right hand near the end of one blade and the left hand near the root of the extended or working blade. The palms of both hands face downward against the paddle. One of the subtle but critical points that the kayaker must remember in capsizing for this and the other sculling rolls is that once the movement has begun, one must try to complete the roll without losing momentum. Since the direction of the capsize is away from the extended paddle blade, the blade is moving fast when it hits the water surface on the other side, after moving in a 180-degree arc up and over the kayak. Thus, it is important that the extended blade be turned on edge at the beginning of the capsizing phase, or else the flat blade will slap

the water and the momentum will be lost. Keeping the extended blade on edge as it strikes the surface allows it to slice into the water. Then the kayaker immediately changes the blade angle by flicking the wrists and sculling to keep the kayak rotating.

The sculling movement required for this roll is considerably more vigorous than that required for the easiest sculling paddle brace, *innaqatsineq* (sculling on the back). In the latter maneuver the kayak is kept heeled over with the lower knee so that it tends to fall away and the kayaker uses the immersed portion of the body for buoyant support. Very little work is required to sustain the kayaker's position on the surface. However, more effort is required to rotate kayak and kayaker through the last half of this sculling roll. It helps if one keeps the kayak rolling smoothly throughout the maneuver to maintain momentum. This maneuver was the eighth one that Crantz described in his list of rolling methods observed more than two centuries ago. Crantz gave the reason why this roll is practiced: "This is of service when they lose the oar during the oversetting, and yet see it swimming over them, to learn to manage it with both hands from below" (Crantz 1767, vol I:152–153). One reason why this is the most popular of the sculling rolls is that the method of holding the paddle horizontally under the kayak with the arms extending downward automatically positions the kayaker close to the deck and keeps the kayak and kayaker together during the roll, thus keeping the kayaker oriented and reducing the force needed to right the kayak.

Sculling Roll with Paddle across *Masik* (*Masikkut Aalatsineq*). Based on my observations in Greenland, this sculling roll is perhaps the second most popular of Greenland sculling rolls, after the *qaannaq ataatigut ipilaarlugu*. The name *masikkut aalatsineq* means "to hold the paddle across the *masik*"—the wide, curved deck beam that supports the forward edge of the cockpit hoop—"and to scull the paddle back and forth." To practice this maneuver, the paddle is held across the *masik* with one end extended to the side away from the direction of the capsize. Thus, for a clockwise roll the paddle will be extended to the left and the kayaker will capsize to the right while grasping the paddle with the hands near the gunwales on each side of the kayak. The right hand grasps the paddle blade near the tip end and the left hand grasps the blade near the root. As the kayaker capsizes to the right, the same precaution of having the blade enter the water on edge applies as described above for *qaannaq ataatigut ipilaarlugu*. If the paddle blade hits the water with the flat face, it tends to stop the roll. So the kayaker flicks the wrists to make the paddle slice into the water surface blade-on-edge, then immediately changes the blade angle to the flatter sculling angle and sculls vigorously to keep rolling before the momentum of capsizing is lost.

This roll is very similar to *qaannaq ataatigut ipilaarlugu*, except that the paddle is held across the foredeck instead of under the kayak. Holding the paddle across the foredeck makes it somewhat more difficult to keep control than is the case with sculling underneath the kayak. This is because the kayaker is not as securely united with the kayak when sculling above the foredeck, nor can he or she lean as far forward.

Steve Burkhardt, an American kayaker skilled in Greenland kayaking technique, has found that it is easier to transmit the force of sculling to the kayak when holding the paddle across the bottom of the kayak than when holding the paddle across the deck. The higher degree of difficulty probably explains why this roll is second to *qaanap ataatigut ipilaarlugu* in popularity. Yet one Greenlander appears to do both maneuvers with equal ease. John Petersen, champion in three successive Greenland National Kayaking Championships, performs *masikkut aalatsineq* by holding the paddle firmly against the foredeck with his hands far enough outboard the gunwales to prevent the knuckles from hitting the deck during the sculling movement. He leans well forward as he capsizes to keep his body as close to the foredeck as possible and sculls around in three or four vigorous back-and-forth sculling movements. Petersen told me in 1995 that he finds it easier to scull with the paddle across the deck instead of holding it under the kayak.

Vertical Scull (*Qiperuussineq / Paatit Ammorluinnaq*). This is a sculling roll in which the paddle is held vertically as deep in the water as the kayaker can comfortably reach. Thus it requires deeper water for practice than any other rolling maneuver. For most kayakers with paddles of typical length, a depth of at least 8 feet (2.5 m) is required, but a deeper depth would be preferred. To practice this maneuver, the kayaker inserts the paddle vertically into the water on the side toward which the capsize will be made. This maneuver is usually practiced by capsizing and recovering on the same side largely because of the need to shift the hand position during the capsize.

To capsize to the right, the paddle, held in the right hand, is passed vertically into the water on the right side of the kayak until only a short part of the vertical blade remains above the surface. The kayaker grasps the blade at the surface and holds it as one would a dagger that was pointed downward. Thus the right hand grasps the paddle blade several inches below the end of the blade, with the palm against the blade and the knuckles facing outboard. The kayaker leans to the right and capsizes. As the capsize progresses, the kayaker's left hand reaches toward the vertical paddle blade to grasp it between the right hand and the tip of the blade. The knuckles of both hands are outboard so that both hands are gripping the same blade as the kayaker's body plunges under the surface to become upside down.

Once the capsize is complete, the kayaker adjusts the hand positions to a distance apart that is comfortable for sculling—about a body width apart—by shifting the hand toward the capsize further down the blade. Then the fore-and-aft sculling motion is begun and the kayaker sculls to the surface on the same side as the capsize while keeping the paddle sculling in the vertical plane. The vertical position of the paddle is maintained until the kayaker surfaces and becomes upright.

Paddle Held behind Back throughout Roll (*Kingup Apummaatigut*). Another sculling roll can be done by holding the extended paddle across the afterdeck with both hands and sculling. This maneuver was observed by David Crantz (1767, vol. I:152) and is the sixth method on his list of maneuvers. This roll is similar to *masikkut aalatsineq*, the sculling roll with the paddle across the *masik*, except that the paddle is held behind the back across the afterdeck. The palms of both hands face the deck, with the thumbs around the forward edge of the paddle. For a clockwise roll, the extended end of the paddle is held outboard of the starboard afterdeck gunwale. After the kayaker capsizes to the right, the paddle is swept forward across the afterdeck and sculled on the left side as the kayaker leans forward.

Sculling on the Chest Used as a Roll (A *Palluussineq* Variation). The Scott Polar Research Institute film archives in Cambridge has some film footage of Gino Watkins using a kayak. He demonstrates a variation of the sculling paddle brace, *palluussineq*, which is normally done by leaning on the water facing outboard with the chest downward. But Watkins leans backward to assume the position used to perform the *palluussineq*, then capsizes backward and sculls around in a complete roll to finish in the *palluussineq* position.

The sculling movement and body position are held throughout the maneuver so that the force generated rotates the kayaker through the water. If the kayaker wishes to pause on the surface before becoming upright, the sculling is continued. To become upright the kayaker increases the opposing hand forces on the paddle and rolls up.

Deck Strap Sculling Roll. The deck straps that hold a seal catcher's hunting implements on a Greenland kayak are made from the tough skin of the bearded seal. As a rule, these straps run straight across the deck at right angles to the longitudinal axis of the kayak. When a kayaker rests at sea, it is customary to run one end of the paddle under the foredeck straps. The free end of the paddle trails aft at an angle of about 45 degrees in order to provide a brace. (This is visually similar to Figure 1.19) Some kayaks have some of the deck straps arranged diagonally, which makes it easier to have the free end of the paddle project at a 90-degree angle to the kayak. A paddle that is extended at a right angle provides more leverage for steadying the kayak than one that is at a 45-degree angle.

With the paddle at either angle, it is possible to roll the kayak by sculling the extended end of the paddle. I believe that the third maneuver on the list of ten maneuvers described by Crantz was a sculling roll: "They run one end of the *pautik* under one of the cross-strings of the kajak, (to imitate its being entangled) overset, and scrabble up again by means of artful motion of the other end of the *pautik*" (Crantz 1767, vol. I:152). Although I have never seen this maneuver performed, it is possible to scull through the 180-degree arc that is the underwater part of the roll. It would reduce resistance if the kayaker leaned forward and wrapped the free arm under the kayak in order to help keep the torso close to the foredeck while the other arm sculled the free end of the paddle.

Figure 1.21—Luutivik Tittussen cod fishing in the fjord near Aappilattoq, Cape Farewell District, Greenland, in a canvas-covered kayak, 1971. Note the angle at which Tittussen holds the paddle. Courtesy Ove Bak.

Reverse Sweep with Paddle behind Neck (*Kingukkut Tunusummillugu*). This maneuver is identical to *kingumut naatillugu*, reverse sweep with paddle across chest, except that the paddle is held behind the neck instead of in front of the body.

Sweep from Bow, Paddle behind Neck (*Siukkut Tunusummillugu*). For a clockwise roll (Figure 1.22), the right hand grips the paddle near one end and the left hand holds it near the middle. The paddle is extended to the left while held behind the neck. Then the kayaker turns the torso to the right and hooks the extended blade across the right gunwale near the bow. The capsize is made toward the right. To recover, the paddle is swept aft and the kayaker leans aft during the sweep. The kayaker finishes the roll lying on the afterdeck with paddle behind the neck, extended to the left.

Rolling with Arms Crossed (*Tallit Paarlatsillugit Paateqarluni / Masikkut*). For this roll (Figure 1.23) the arms are crossed, which puts the kayaker into a strained position during recovery. For a clockwise roll, the paddle is held blade-on-edge along the starboard gunwale with the right hand near the middle of the paddle. The left forearm is held across the body; the left hand grasps the paddle near the aft end, below the forward-reaching right arm. The knuckles of both hands face outboard. The kayaker capsizes to the right. To recover, the paddle is swept aft and sinks deeper than in most rolls as the kayaker emerges, leaning forward as the kayak becomes upright.

Paddle Held Lengthwise along Spine, Hands Held Fore and Aft (*Aariammillugu*). For a clockwise roll, the paddle is held near the end with the right hand low and behind the back and the left hand grasping the paddle behind the kayaker's neck. The paddle angles upward and forward at an angle of about 45 degrees with the deck of the kayak, as seen from the side. The right hand grips the after end of the paddle with the palm against the lower side of the blade, behind the kayaker. The left hand grasps the upper side of the paddle near the middle, with the palm against the paddle. The kayaker leans well forward and capsizes to the right. The recovery is initiated by sweeping the paddle and torso aft, raising the left thigh toward the deck to rotate the kayak. The roll is finished with the kayaker leaning aft.

Paddling Upside Down with Paddle across Keelson (*Pusilluni Paarneq*). Paddling upside down is done for sport at kayaking practice sessions. Sometimes two or more kayakers will have an

Figure 1.22a—Sweep from bow with paddle behind the neck (*siukkut tunusummillugu*). Pavia Tobiassen prepares to capsize. All photos courtesy Vernon Doucette, 1994.

Figure 1.23a—Rolling with arms crossed (*tallit paarlatsillugit paateqarluni/masikkut*). Ove Hansen in starting position for a left-handed roll. All photos courtesy Vernon Doucette, 1994.

Figure 1.22b—Tobiassen keeps the paddle behind his neck until completely capsized. He will sweep his paddle aft to begin recovery.

Figure 1.23b—After capsizing, Hansen starts the recovery by sweeping the paddle aft.

Figure 1.22c—Tobiassen nears completion of the roll with his body and paddle near the afterdeck.

Figure 1.23c—Hansen finishes the roll.

informal race while paddling upside down. This is also an impressive maneuver to perform at exhibitions. It is fairly easy to perform by holding the paddle across the bottom of the capsized kayak and making a back-paddling motion to go forward, because everything is reversed when one is upside down.

Hands in Normal Position, Roll Forward (*Siukkut Pallortillugu / Masikkut*). This roll is performed with the hands in the normal paddling position instead of extending the paddle for maximum leverage. Thus it is similar to the popular methods of rolling used by recreational kayakers today in whitewater rivers. For a clockwise roll, the paddle is held with the hands in the normal paddling position along the starboard side of the kayak. The kayaker leans forward and toward the paddle to capsize on the right. The torso is kept in this position until the kayak is completely upside down. The head and torso are near the surface on the recovery side at the instant the capsize has ended. The timing of the recovery is more critical for this roll than some of the others and it is easier to roll up if the recovery is begun before the capsize momentum is lost. To recover, the forward end of the paddle is swept aft. As the paddle begins to create lift, the left thigh is pressed against the deck beam to roll the kayak as far as possible while the head and torso are still in the water. The kayaker's head and shoulders emerge when the kayak is almost upright. The body is leaning slightly forward as the roll is completed.

Coping Without a Paddle. Greenlanders do not carry a spare paddle on the deck of their kayak. Their hunting technique is so standardized and specialized that almost all of the available deck space is filled with hunting equipment, with each item in a special place. In calm sea conditions, the harpoon or lance can be used as an improvised paddle. Either of these implements can be used as an outrigger to steady the kayak if the paddle cannot be found and the kayaker has to wait for help from a nearby companion. The hunting float can also be used to steady the kayak. If the kayaker happens to be alone, the harpoon or lance might even be used as an emergency paddle to reach shore.

If a paddle becomes lost during a capsize, a skilled kayaker might still roll up with the harpoon or lance by using either implement as one would use a paddle. The hunting float can also be used for rolling up; some Greenlanders practice by using it for a complete roll, as described next.

Roll Using Sealskin Float (*Avataq* Roll). The *avataq*, or hunting float, is made from the entire skin of a small seal, which has been care-

fully skinned in such a way as to leave the skin as intact as possible. Any openings in the skin are carefully sewn to make an air- and watertight unit. A spool-shaped mouthpiece and plug are fitted to the float in order to inflate or deflate it. At one end of the float is a short line that attaches to the harpoon line. At the opposite end is a loop. By running the line through the loop, a handle is formed that provides a place for a kayaker to grasp the float near each end. In an emergency, the float can be grasped in whatever way the kayaker can hold it and used to roll up or at least get the face above the surface and get a breath of air. To practice a complete roll, it is necessary to let some of the air out of the float so that it can be forced underwater for rolling.

For the float roll (Figure 1.24), the kayaker grasps the strap so that the fists are against the float and the heel of each hand is at the ends of the strap. For a clockwise roll (Figure 1.25), the float is held lengthwise along the right side of the kayak, with the right hand near the kayaker's right hip and the left near the right thigh. The kayaker leans over the float and forces it under by pushing downward with both fists, keeping the float right against the side of the kayak until the capsize is completed.

When the kayak is fully capsized, the float will pop to the surface on the opposite side (Figure 1.25b). To become upright, the kayaker pulls the left, or forward, hand downward to start the recovery. This makes the float tip down at the front end as the kayaker's weight is put on it. The left, or trailing, thigh is pressed against the deck beam to rotate the kayak ahead of the kayaker. As the float sinks, the right hand, which is holding the aft end, flips upward and forward. The kayaker leans aft as the float passes the torso in order to let the float

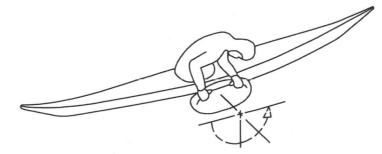

Figure 1.24—Roll using a sealskin float (*avataq* roll). The straps on a partially inflated float are tied together so that the float can be held firmly. For a clockwise roll, the float is held near the right hip and the kayaker pushes the float under to capsize. The position shown is held until the kayak is upside down and the float surfaces on the far side. The kayaker uses the buoyancy of the float to roll himself up. The roll ends with the float near the left hip of the kayaker.

Figure 1.25a—*Avataq* roll. Ove Hansen holds the sealskin float along the starboard gunwale. All photos courtesy Vernon Doucette, 1994.

Figure 1.25b—When fully capsized, the float pops to the surface on the opposite side. To start the recovery, Hansen pulls down on the float with his left hand.

Figure 1.25c—As Hansen surfaces, he holds the float away from the kayak to increase leverage to right the kayak.

Figure 1.25d—As he finishes the roll, Hansen's right hand is forward, left hand is aft.

clear and also to keep the weight near the kayak. However, it may help to hold the float some distance out from the kayak in order to make more leverage for righting the kayak. As the roll progresses, the float flips forward, so that at the finish the position of the hands and float is a mirror image of the starting position—that is, the float is now on the kayaker's left, with the left hand near the left hip and the right hand near the left thigh or left knee.

THROWING STICK AND HAND ROLLS

Some Greenlanders are able to roll with the *norsaq*, or harpoon throwing stick, and some are able to roll with the hand alone. The throwing stick is tapered, so it is usually held by the small end and used as a one-handed paddle. But I have also seen it held by the large end or even in the middle. The throwing stick rolls that are used in the Greenland Kayaking Association contests will be described here. Since most right-handed kayakers would prefer to roll with the right hand, the throwing stick and hand rolls described will be for counterclockwise rolls.

Throwing Stick Swept Outward and Down from near *Masik* (*Norsamik Masikkut*). For a counterclockwise roll, the right hand, which holds the throwing stick, is laid on the foredeck at an angle pointed forward and to the left. (Figure 1.26 depicts a clockwise roll.) The kayaker leans low over the foredeck and places the free, or left, arm under the kayak. The capsize is made to the left. As the capsize is completed, the throwing stick is quickly swept out to the kayaker's left, then downward, bringing the kayak upright in a continuous movement.

Throwing Stick, Leaning Aft (*Norsamik Kingukkut*). For a counterclockwise roll, the kayaker holds the throwing stick in the right hand and turns the torso to face right, at the same time reaching aft with the right hand and hooking the throwing stick over the port afterdeck edge. Then the kayaker capsizes backward. When fully capsized, the kayaker sweeps the throwing stick forward to roll the kayak. As the kayaker emerges, he or she throws the free (left) hand over the kayak as a counterbalance.

Throwing Stick Aft, Then Forward (*Norsamik Nerfallaallugu*). For a counterclockwise roll, the kayaker holds the throwing stick in the right hand on the foredeck, pointed forward and to the left. The capsize is to the left. After capsizing, the throwing stick is swept aft as far as the kayaker can reach, then forward, taking additional sculling strokes if necessary to get upright. The free (left) arm is thrown

Figure 1.26a—*Norsamik masikkut.* The throwing stick (*norsaq*) is swept outward and down. Photo shows a left-handed roll. All photos courtesy Vernon Doucette, 1994.

over the kayak as the kayaker emerges. The kayaker finishes lying back on the afterdeck, arms down.

Hand Only, Sweep Outward and Down from Near *Masik* (*Assammik Masikkut*). Same as *norsamik masikkut* (see above), except the right hand is used for rolling (Figure 1.27) instead of the throwing stick.

Hand Only, Leaning Aft (*Assammik Kingukkut*). This maneuver is the same as the *norsamik kingukkut* (see above), except the right hand, not the throwing stick, is used for rolling. A variation is to turn the body leftward during the capsize, reaching aft as far as possible on the port side as the kayak capsizes. Then the forward sweep is made as in the *norsamik kingukkut*.

Hand Only, Sweep Aft, Then Forward (*Assammik Nerfallaallugu*). Same as the *norsamik nerfallaallugu* (see above), except the right hand, not the throwing stick, is used for rolling.

Clenched Fist Roll (*Assak Peqillugu, Qilerlugu / Poorlugu*). In the clenched-fist roll, the kayaker sometimes holds a rock in the hand to prove that the fist remains clenched. The movement is the same as the *assammik masikkut* (hand only, sweep outward and down from near *masik*), except that the fist must move very quickly and the hip snap that rotates the kayak must be vigorous.

Rolling with a Rock Swept Outward and Down (*Ujaqqamik Tigumisserluni*). In rolling with a rock, the movement is the same as the *assammik masikkut* (hand only, sweep outward and down from near *masik*), except the kayaker's right hand holds a sizable rock for

Figure 1.26b—Hansen capsizes to the right, keeping the *norsaq* against the foredeck in his left hand and his right hand on the bottom of the kayak.

Figure 1.26c—Starting the recovery.

Figure 1.26d—Hansen sweeps the *norsaq* back and uses his right arm to counterbalance.

Figure 1.26e—Hansen finishes the roll lying back on the afterdeck, arms down.

Figure 1.27a—*Assammik masikkut*. Hand only, sweep outward and down from near the *masik*. Tobiassen prepares to roll with his right hand forward, left hand holding the hull.

Figure 1.27b—When fully capsized, Tobiassen sweeps his right hand outward and down.

Figure 1.27c—Tobiassen finishes the recovery using his left arm to counterbalance.

a counterclockwise roll. At Sisimiut in 1995, a 9.7-pound (4.4 kg) concrete paving block was used. The movement must be very quick, according to three-time national champion John Petersen.

Elbow Roll with Hand Held against Neck (*Ikusaannarmik Pukusuk Patillugu*). For a clockwise roll, the kayaker leans forward and to the right with the elbow bent so that the left hand rests on the nape of the neck (Figure 1.28). The kayaker capsizes to the right. To recover, the head, hand, and elbow are swung outward and aft as the left thigh is pressed against the deck to rotate the kayak. The free (right) arm is thrown over the kayak as a counterbalance as the roll is completed with the kayaker leaning aft.

Straightjacket Roll (*Tallit Paarlatsillugit Timaannarmik*). In 1985, Manasse Mathaeussen showed me how this roll was done. He had been able to do it from the time he was 16 years old until he was 60 years old, but he was then 70 and could no longer do it. For a counterclockwise roll, the kayaker crosses the arms as if in a straightjacket and leans well forward. The capsize is made to the left. The instant that the capsize is complete, the kayaker, with arms still crossed, leans aft as quickly as possible, keeping the torso twisted to the left to act as a hydrofoil. If the hands can be released from the straightjacket position just as the kayaker surfaces, it makes it easier to get around the edge of the kayak and keep it under the kayaker. The face is turned forward as the head surfaces at the left afterdeck edge. The left ear and knuckles of both hands push the left afterdeck edge as the kayaker surfaces.

Mathaeussen died in 1989. In 1992, I learned that no one in Greenland knew how to do this roll. So even though I was not flexible enough to do this roll myself, it was possible to show the movement to John Petersen. By 1995, at least five kayakers in Greenland could do it. Thus Mathaeussen was able to pass this roll along to others more than two years after his death. In 1995, the contest judges required contestants to keep the arms crossed until the end of the roll, which makes the maneuver very difficult.

The maneuvers described above include most of the 30 maneuvers on the list used in the Greenland Kayaking Association competitions. The three maneuvers listed below are based on others on the list:

> Maneuver 16: *Pallortillugu assakaaneq 10 sekuntit*. This is basically *siukkut pallortillugu masikkut* (hands in normal position, roll forward, as described above), except that the kayaker does as many as possible in ten seconds (before 1995, it was five rapid rolls without the time limit).

Figure 1.28—The elbow roll. The solid lines represent the starting point for a roll to the right (1). The dashed lines represent the finishing position (2). The curved arrow is the path that the kayaker's elbow will follow relative to the kayak.

To prepare for rolling, the kayaker puts the palm of the left hand against the nape of the neck and twists the torso so that he or she faces to the right. The left elbow points forward and is held near the foredeck. The right hand is held against the right side of the kayak. The kayaker leans low and to the right to capsize.

The kayaker allows the momentum of the capsize to bring the body just past the fully capsized position before starting recovery. Timing is critical—the kayaker must initiate recovery before losing momentum. He or she straightens the torso and brings the left elbow outboard and aft. The right hand is thrust outward as a counterbalance. The speed that is required to complete this roll and the sudden untwisting of the torso while bending backward create a risk of injury.

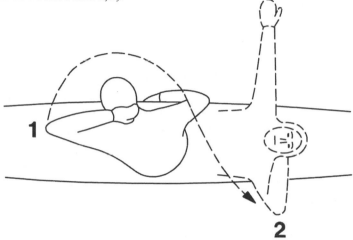

Maneuver 17: *Nerfallaallugu assakaaneq 10 sekunit*. This is basically *kinnguffik paarlallugu* (the standard Greenland roll, described above), except that the kayaker does as many as possible in ten seconds (before 1995, it was five rapid rolls without the time limit).

Maneuver 30: *Nusutsinneq*. This is an application of the brace *palluussineq* (sculling on the chest, described above) used in competition by having five men drag a kayak sideways. They place a stick under the deck straps behind the kayaker and fasten a line to it. The line is pulled to simulate a large seal dragging a kayak, as may happen if the hunting float did not get thrown free of the kayak after a large seal had been harpooned. This is a dangerous simulation, because the kayaker's head is dragged underwater and he cannot always signal for the men to stop pulling on the line if something goes wrong.

One of the rolls that Manasse Mathaeussen used to do was to roll with his mittened hands held together. Mathaeussen also did "showmanship" rolls, such as switching paddle positions underwater or rolling with his little finger. John Petersen, the 1995 champion, often does the cigarette trick, in which a cigarette is kept above the water in one hand as the kayak is capsized, then passed over the bottom of the capsized kayak to the other hand before the kayaker rolls up with the hand that had held the cigarette. Although such maneuvers have no practical value, they do help maintain public interest in Greenland's kayaking heritage.

Of the 40 or so capsizing maneuvers discussed above, the first few are the most important ones to learn, so they were described in more detail. But most Greenlanders cannot roll at all. One of the veteran seal catchers at Sisimiut in 1995 could not do any of the capsizing maneuvers that the youngsters were performing. But he had once caught 20 seals in one day, which won him more respect in his community than he would have gotten as a champion kayaker.

MISCELLANEOUS REMARKS

There are a few items in my field notes that are too brief to use as a chapter, yet too interesting to ignore. They have been saved for this final section because they furnish a glimpse of Greenland's kayaking heritage.

The first is a legend that has been handed down from generation to generation. In 1992, Gail Ferris and I were the only two non-Greenlanders at a kayak training camp at the abandoned village of Ikamiut (65°37.5' N, 52°46.9' W), which is located on Tuno (Hamborgersund), about 15 miles north of Maniitsoq. Tuno is an elbow-shaped body of water between Hamborger Island (Hamborgerland or Sermersuut) and the mainland. Mount Ingik (65°37.5' N, 52°51.5' W) is located on Hamborger Island at the bend of the elbow, directly across from Ikamiut. While we were at Ikamiut, Kâlêraq Bech, past president of the Greenland Kayaking Association, pointed out that missionary Hans Egede had preached a sermon at the foot of Mount Ingik in the 18th century. At that time, Ikamiut was a flourishing village, and the local people crossed Tuno in umiaks and kayaks to hear Egede's sermon (Egede 1741).

Then Bech pointed to the top of Mount Ingik and told the legend of Parnuna. Parnuna lived at Ikamiut and her husband was a seal catcher. One day, her husband went out in his kayak on Tuno. That evening, Parnuna began looking for him because he had not returned. Finally, she decided to climb Mount Ingik in order to get a better view of Tuno. From the top of Mount Ingik, Parnuna saw her husband's overturned kayak far below. Realizing that he was dead, Parnuna leaped to her death off the steep face of the mountain. No one knows if the story of Parnuna is true, but almost everyone in the area has heard it. The tale underscores the risks of seal catching and the anxious vigil that many wives have endured.

During the final celebration of the 1995 kayaking championship at Sisimiut (Map 3), I received an invitation to visit the mayor's office. Mayor Simon Olsen wanted to talk about kayaks, and we had a pleasant discussion of almost two hours, with the city manager interpreting from Danish to English. The mayor was not a kayaker, but older members of his family had been. One of the stories he had been told as a youngster was how the seal catchers would protect themselves from killer whales. Seal catchers would often tow large seals beside the kayak and carry small ones on the afterdeck. This might attract killer whales, especially if the seal happened to be trailing blood; so if killer whales were in the area, a kayaker would tie a knife on a line and hang the knife one or two meters below the kayak. According to the mayor, this would keep killer whales away.

I discussed this method with Kâlêraq Bech. He believed that the flashing of the knife blade as it hung below the kayak tended to frighten the killer whales. When I discussed Mayor Olsen's story with veteran seal catcher Karl Samuelsen, Samuelsen said that he preferred to use his *unaq*, or large harpoon, to scare killer whales away. He would thrust his *unaq* vertically into the water so that the foreshaft was about three feet (one meter) below the surface. The white foreshaft apparently frightens the killer whales away from the kayak.

Another danger that kayakers sometimes encounter was related to me in 1961 by Zachary Gussow of Louisiana State University. He said that Greenlanders are the only Native kayakers to report the nervous disorder known as *kajakangst*, Danish for "kayak anxiety." It is also known as "kayak phobia" or "kayak fear." This condition usually begins with a dizziness that the Danes call *svimmelhedden*, which occurs when the kayaker is alone on a calm, reflective sea. The kayaker might lose his balance and become afraid that he would capsize and drown. The condition can disappear at the approach of other kayakers. Sometimes the kayaker develops bizarre symptoms. For example, he might believe that his kayak was filling with water, growing longer, and so on.

In 1985, while doing research for a magazine article on *kajakangst*, I contacted NASA physicians Walter Davis and Heidi Kapanka. After hearing the description of the kayaker's unsteady feeling, they explained that it was similar to that experienced by airplane pilots or astronauts when they lose a visual reference point. They called it spatial disorientation (Heath 1986).

I have experienced spatial disorientation while in a kayak, but it was while near shore and it disappeared after finding a reference point. From the Greenland descriptions of the feeling going away when other kayakers approach, it is assumed that the approach of other kayakers helped the victim orient himself. However, lack of a reference point does not completely explain the psychological aspect of *kajakangst*.

In 1993, a psychiatrist who was also a kayaker became interested in *kajakangst*. Lynn Hunter Hackett and her colleague, Mary Beth Shea, viewed some Danish studies of Greenlanders who had experienced *kajakangst*. Dr. Hackett gave her preliminary opinion: "The case histories at this point describe symptoms utterly consistent with the diagnosis of panic disorder, and then the development of a debilitating agoraphobia (Lynn Hunter Hackett, letter, May 13, 1993).

In closing, I believe that my tremendous admiration for all Inuit kayaks was best expressed by Greenlander Kâlêraq Bech. He was, at the time, chairman of the Greenland Kayaking Association. His speech at the 1995 competition was in Greenlandic and a portion is paraphrased here.

THE GREENLAND KAYAK The Greenland kayak is the ingenious result of using the minimal resources available to the Greenlanders, who found the best way to overcome the power of nature and successfully hunt sea mammals in a very hostile part of the world.

A Greenland kayak is built to use out in the open sea under any and all arctic conditions. The shape is adapted to the owner in every way. This design and adjustment has been going on for centuries through the experiences of the seal catchers. The attempts of outsiders to alter the kayak's unique technical features are likely to have a harmful effect.

A Greenland kayak is not designed to be abandoned in emergency situations and is built as a rescue craft. Its narrow width, low height, and the fitting of the cockpit to the owner make it easy both to propel and to transport. The clever design and construction allow the hunter to maneuver the kayak easily and to fight storms. The kayaker can recover elegantly if he capsizes.

While the kayak may at first appear to be a fragile craft, each part of the construction combines to make the kayak a strong and unique vessel. This can be confirmed by the seal catchers, who, to get nearer the seals, waited until it was stormy. In order to gather experience in rough seas, they practiced on the breaking waves among the small islands, rocks, and reefs. Another example of the strong construction of the kayak is seen when seal catchers were caught in a storm and forced to go ashore. At times, the only way to get home was to get into the kayak on top of a steep rock, then seal the *tuilik* (kayak jacket) to the kayak. When the best part of a wave was under them, they let the kayak drop into the water from steep slopes with heights of several yards (several meters). You can imagine the force

of a fully grown man of 155 to 175 pounds [70 to 80 kg] landing in the water. This is proof of excellent design and construction.

The kayak is so strong because of the construction, not in spite of it. Ribs fitting into carved holes and slender sticks lashed together and covered with a tough, resilient skin are essential features, not ignorance. Modern materials such as nails are simply inadequate.

The seal catchers do not worry about the construction of the traditional Greenland kayak when hunting even in the strongest storms. As an old seal catcher once said, "I have trust in my vessel in all kinds of weather; it is only me who feels weak."

REFERENCES

Anonymous

1934　The Eskimo Kayak. *Polar Record* 7:52–62.

Birket-Smith, Kaj

1924　Ethnography of the Egedesminde District with Aspects of the General Culture of West Greenland. *Meddelelser om Grønland* 46.

1953　*The Chugach Eskimo*. Copenhagen: Nationalmuseets publikationsfond.

Brand, John

1984　*The Little Kayak Book*, part I. Colchester, Essex: John Brand.

Chapman, F. Spencer

1934　*Watkins' Last Expedition*. London: Chatto and Windus.

Crantz, David

1767　*The History of Greenland: Containing a Description of the Country and Its Inhabitants and Particularly a Relation of the Mission*, 2 vols. London: Brethren's Society for the Furtherance of the Gospel.

Egede, Hans

1741　*Det Gamle Grønlands Nye Perlustration eller Naturel-Historie*. Copenhagen: J. C. Groth.

Fabricius, Otto

1962　Otto Fabricius' Ethnographical Works. Erik Holtved, ed. *Meddelelser om Grønland* (Copenhagen) 140(2).

Gabler, Ulrich

1982　Technisches Gutachten. *Mitteilungen der Geographischen Gesellschaft zu Lübeck* 55:226–230.

Gosch, C. C. A., ed.

1897　*Danish Arctic Expeditions, 1605 to 1620*, 2 vols. London: Hakluyt Society.

Gronseth, George

1992　Learning to Kayak the Greenland Way. *Sea Kayaker* 32(Spring): 46–55.

Heath, John D.

1978　Some Comparative Notes on Kayak Form and Construction, pp. 19–26 in D. W. Zimmerly, ed., *Contextual Studies of Material Culture*, National Museum of Man Mercury Series, Canadian Ethnology Service Paper 43, Ottawa.

1986　*Kajakangst*: The Greenland Hunter's Nightmare of Disorientation. *Sea Kayaker* 2(4):62–64. [A revised version appeared in 1991 in E. Y. Arima et al., eds. *Contributions to Kayak Studies*, pp. 93–106. Canadian Museum of Civilization Mercury Series, Canadian Ethnology Service Paper 122, Hull.]

1987　The Phantom Kayakers, a Scottish Mystery. *Sea Kayaker* 4(1): 15–18.

1992　The Balance Brace. *Sea Kayaker* 8(4):56–58.

MacRitchie, D.

1912　The Kayak in North-Western Europe. *Journal of the Royal Anthropological Institute* 42:493–510.

McGhee, Robert

1984　Thule Prehistory of Canada, pp. 369–376 in D. J. Damas, ed., *Arctic*, Handbook of North American Indians, volume 5. Washington, D.C.: Smithsonian Institution.

Mikkelsen, Ejnar

1954　Kajakmanden fra Aberdeen. *Grønland* (Copenhagen), pp. 53–58.

Nooter, Gert

1971　Old Kayaks in the Netherlands. *Mededelingen van het Rijksmuseum voor Volkenkunde* (Leiden) 17.

Petersen, H. C.

1982　*Qaanniornermut ilitsersuut. Instruktion i kajakbygning. Instruction in Kayak Building*. Trans. G. Fellows Jensen. Roskilde: Greenland Provincial Museum and the Viking Ship Museum.

1986　*Skinboats of Greenland*. Trans. K. M. Gerould. Ships and Boats of the North, vol. 1. Roskilde: National Museum of Denmark, the Museum of Greenland, and the Viking Ship Museum.

Reid, R. W.

1912　Description of Kayak Preserved in the Anthropological Museum of the University of Aberdeen. *Journal of the Royal Anthropological Institute of Great Britain and Ireland* 42:511–14.

Souter, W. Clark

1934　*The Story of our Kayak and Some Others*. Aberdeen: University of Aberdeen Press.

Wallace, James

1700　*An Account of the Islands of Orkney*. London: Jacob Tonson.

Whitaker, Ian

1954　The Scottish Kayaks and the 'Finn-men.' *Antiquity* 28:99–105.

2 Using Greenland Paddles: An Overview

GREG STAMER

Interest in Greenland kayaking technique in the United States and Canada has grown in recent years. This is due, in part, to the availability of excellent new videos and to a growing body of literature and instruction. Greenland kayakers such as Maligiaq Padilla, who has won five of the last six Greenland National Kayaking Championships, are traveling globally to teach. In addition, the Greenland Championships have been opened to outsiders, allowing participants to learn Greenland technique firsthand.

My own understanding of Greenland technique was initially forged from discussions with kayak scholar John Heath and from intense study of his early kayaking videos. In recent years I have enjoyed the opportunity to apprentice with and teach alongside Maligiaq Padilla, past president of the Greenland Kayaking Association (Qaannat Kattuffiat) Kâlêraq Bech, past president of Qajaq Nuuk and Qajaq København Pavia Lumholt, and other fine kayakers. Certainly my most influential experiences are those that have occurred in Greenland, including attending a Greenland Kayaking Association training camp as an invited student in 2000 and participating in the Greenland National Kayaking Championships in 2000 and 2002. The Greenland training camps and competitions are remarkable environments for learning, as the participant is surrounded by many of Greenland's finest kayakers, including retired seal catchers.

The following information on stroke styles and stroke mechanics is not meant to be definitive, since space limitations do not allow for a comprehensive treatment of the subject and I still consider my understanding of various Greenland kayaking techniques to be incomplete. In addition, some of the techniques that I have been exposed to in competition may differ from techniques commonly used for hunting. Greenland kayaking technique has many regional and personal variations. Every time I have had the opportunity to work closely with Greenland kayakers, I have learned new concepts. My forward paddling stroke with a Greenland paddle continues to develop over time through small, incremental improvements, occasional dead-end experiments, and intermittent epiphanies.

INTRODUCTION

In early summer 2002, I sat on the ice-choked shore of Ilulissat, Greenland, and watched as a very experienced Greenland competitor borrowed a contemporary feathered kayaking paddle for the first time. Although he was well versed in about 30 different methods of rolling, the kayaker's habitual grace suddenly vanished. After a short period of time during which the paddle fluttered, or the blade sliced ineffectively into the water, he abruptly ended the trial amid a flurry of good-natured laughter from shore.

I have witnessed a similar event many times; however, in all my prior experiences this transpired when paddlers experienced with using feathered, spooned, touring paddles tried a Greenland paddle for the first time. An important lesson to be learned for kayakers experimenting with different paddle types is that although there is significant overlap in fundamentals, many subtle differences in technique must be learned in order to use the equipment to its fullest potential. Expect the process to take some time.

Greenland paddles appear quite different from contemporary sport paddles (Figure 2.1). The blades of Greenland paddles are quite narrow and set in the same plane (unfeathered). When swept

Figure 2.1—Paddle profiles. Top to bottom: Epic Mid Wing, adjustable length/feather, blade width 6 ½ in. (16.4 cm), blade length 19 ¾ in. (50 cm); Werner Kauai spooned dihedral, length 86 ½ in. (220 cm), blade width 7 ¼ in. (18.5 cm), blade length 18 ½ in. (47 cm); Superior Kayak carbon fiber shouldered Greenland paddle, length 86 in. (218.4 cm), blade width 3 ½ in. (8.9 cm), blade length 33 ¼ in. (84.5 cm); Western red cedar (homemade) shouldered Greenland paddle, length 87 ½ in. (222.3 cm), blade width 3 ⅛ in. (7.9 cm), blade length 32 ¾ in. (83.2 cm); Superior Kayak unshouldered Greenland paddle, length 86 in. (218.4 cm), blade width 3 ½ (8.9 cm), blade length 32 ¾ in. (83.2 cm); "Storm paddle" by Gabriel Romeu, length 72 in. (182.9 cm), blade width 3 ⁹⁄₁₆ in. (8.8 cm), blade length 31 in. (78.7 cm). Photo by Greg Stamer.

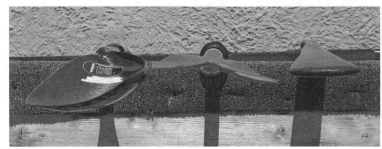

Figure 2.2—Blade tip profiles. Left to right: Epic Mid Wing, Werner Kauai spooned dihedral paddle, homemade Greenland paddle. Photo by Greg Stamer.

through the water with the leading edge elevated, these blades generate a surprising amount of lift, with little tendency to dive. This makes them very effective tools for rolling and sculling. The blades are symmetrical, unlike the curved or spooned shapes of many contemporary touring paddles (Figure 2.2). This design enables the paddle to perform consistently and predictably whether used with a high or low brace or swept forward or backwards. Greenland paddles are custom-sized to permit the kayaker to comfortably grasp the paddle at any point, including the blades. For braces, sweep strokes, and rolling, the kayaker may slide his or her hands and extend the paddle to any degree necessary to maximize leverage and versatility. In addition to paddle extension, sliding the hands on the blades can also limit the length of the paddle that is outboard of the kayaker's body into the wind, thus reducing wind resistance. Finally, Greenland paddles tend to be very buoyant as compared with most synthetic sports paddles, enhancing forward strokes as well as rolling, sculling, and static brace techniques.

PADDLE FEATHER

Greenland paddles have both blades set in the same plane (unfeathered), while the majority of contemporary sports paddles have the blades set at angles to each other (feathered). The question of feathered versus unfeathered paddles is a cherished debate in paddling circles, with vocal proponents of both styles. Each style is a compromise.

Some proponents of feathered paddles claim that feathering gives them slightly more reach and more efficient body mechanics as well as less wind resistance in head winds. When using a feathered paddle, one hand acts as the control hand (the right hand for a right-hand control paddle), and the blade opposite the control hand must be rotated into perpendicular orientation in order to take a stroke. Although wrist problems such as tenosynovitis (inflammation of the tendon sheath) are often attributed to feathered paddles, with proper technique and a relaxed grip, much of this rotation can be accomplished without bending the wrist, using a combination of elbow lift and forearm rotation.

The feather angles of sport paddles continue to change over time. While 90-degree feather angles were popular ten or more years ago and provide the least wind resistance, they also require the most blade rotation. Most feathered paddles sold today have 60 degrees of feather, as a compromise, and specialized paddles for whitewater "rodeo" play can have 15 degrees or even less feather. Since the strokes are asymmetrical, many kayakers using feathered paddles speak of an "on-side" and an "off-side" for rolling and bracing. While kayakers using Greenland paddles must also train on each side for skills such as rolling, the movements are symmetrical; some kayakers find the distinction between on-side and off-side to be somewhat blurred.

During use, the airborne blade of a feathered paddle slices edge-wise through the air to lessen resistance in head winds. However, in extremely strong or gusty head winds, feathered paddles (especially those with low feather angles) have a tendency to spin in the kayaker's hands, as the wind deflects off the rapidly changing blade angles, causing one blade to dive and the other to climb. In similar conditions, the Greenland paddle is neutral and easy to control, although the narrow blades can produce noticeably more wind resistance than a feathered paddle. The use of a very short Greenland paddle designed for the sliding stroke produces little wind resistance because there is very little blade projecting outboard of the kayaker's body and into the wind.

In very strong and gusty beam winds—winds hitting the kayak broadside—the horizontal airborne blade of a feathered paddle can prove a liability, as it can be lifted by the wind and potentially cause a capsize. If caught by surprise, a kayaker may need to let go of the paddle with his or her windward hand, enabling the paddle to flop over the kayak. An experienced paddler can avoid this hazard by leaning into the wind and keeping the blade low on the windward side. In contrast, Greenland paddles are very neutral and well behaved in heavy winds from all directions and so cause few surprises.

A unique feature of an unfeathered paddle is that, since both blades are in the same plane, it is well suited for performing as an outrigger to stabilize the kayak. Those who use the kayak for hunting and fishing, as well as touring kayakers, will appreciate this feature. For resting at sea, I will sometimes place one blade across the cockpit coaming and simply lean forward, pressing the paddle blade downward with the weight of my forearms, and lean slightly toward the side of the extended blade. In Greenland, the paddles are often jammed under strong sealskin thongs on the deck to the same effect. This provides a comfortable brace and frees both hands for tasks other than paddling. According to H. C. Petersen, "Arctic kayakers have considered the stabilising effect of the paddle and its use when capsizing more important than the comparatively great wind resistance that parallel blades create" (Petersen 1986:68).

Kayakers can become skilled in the use of both feathered and unfeathered paddles. The main danger of switching between them is that in an emergency, when forced to brace by reflex, the paddle blade may slice water and potentially cause a capsize if the kayaker's bracing technique does not complement the feather angle used. I originally learned how to paddle feathered, and I had to unlearn many ingrained habits to use a Greenland paddle. I now find the symmetry of strokes to be very satisfying and, through practice, my instinctive brace is now tuned for an unfeathered paddle.

PADDLE SIZING

Paddle sizing may seem an odd topic for an overview of paddling technique, but proper sizing is critical to proper technique. A common misconception is that sizing is more important for an expensive, high-tech carbon sports paddle than for a "simple" Greenland paddle made of wood. However, a Greenland paddle, even if made from discarded lumber, demands proper sizing—in some aspects, even more so than carbon sports paddles. In the case of a Greenland paddle, the kayaker's hands must be able to comfortably grasp the blades and the paddle shoulders. The shoulders (if present) dictate hand spacing and require a precise fit. If the kayaker has spent a great deal of time determining the proper shaft length, paddle feather, and blade size for a sports paddle, he or she may likewise require a similar experimentation period to discover the correct properties for a Greenland paddle. A paddle must not only be custom-fitted to each individual's body measurements but also take into account attributes such as the kayaker's strength, the beam and volume of the kayak, and the type of paddling that is done. For example, in Greenland there are differences between paddles used for hunting and those for competition.

PADDLE LENGTH

Historically, Greenland paddles were found in a wide variety of sizes; local variation continues to this day. The short paddles were commonly found in areas frequented by high winds (Petersen 1986:65–66).

Harvey Golden, who has researched approximately 80 Greenland paddles spanning more than four hundred years, writes:

> The longest (non-Polar Greenland) paddle I have seen is 95¼ inches [242 cm] from "West Greenland"—exact locale unknown. The shortest adult paddle I have seen is 76⅝ inches [195 cm] long, again, West Greenland, again, locale unknown. There can be no question that longer and shorter ones existed, sure as the odds against the very longest and very shortest having both ended up in museums. Anthropometric "rules" were likely very localized, i.e., fjord-to-fjord, and were also likely very flexible, subject to personal preference (Golden 2003).

Maligiaq Padilla advises students to think of anthropometric measurements as general guidelines only. The kayaker may need to make several paddles before discovering the best dimensions for his or her particular requirements. As a case in point, when Padilla was my houseguest and was required to make a new paddle to replace his, which had been broken in transit, he did not use direct body measurements. He simply measured his new paddle against his old and added a little length to each blade as an experiment, to increase the "bite" for racing.

Anthropometric sizing methods vary both regionally and temporally. Some anthropometric methods for determining Greenland paddle lengths are:

- The length of one arm span, plus the distance from the elbow to the fingertips. This is for a longer paddle for general touring (Heath 1989).
- Sized so that the kayaker can just barely curl his or her outstretched fingers over the top of the paddle (according to John Heath).
- The length of one arm span plus two hand spans (according to Aqqaluk Heilmann of Upernavik in 2002).
- The length of one arm span, plus the distance from the wrist to the fingertips. This results in a slightly shorter paddle than the methods above, preferred by some more accomplished kayakers or for rolling (Heath 1989).
- The length of one arm span, plus the length of one's underarm (Kangaamiut measurement) (Petersen 1986:64).
- For a short paddle used with a full sliding stroke, a length of only one arm span (according to John Heath).

In my case, the first three sizing methods yield almost the same measurement of 88 inches (223.5 cm), the length that I currently prefer—for reference, I am 5 feet 10 inches (177.8 cm) tall with an arm span of 73 inches (185.4 cm).

PADDLE-SHAFT AND SHOULDERS

Greenland paddles are either "shouldered" or "unshouldered." A shouldered paddle has a noticeable or abrupt transition where the paddle-shaft meets the roots of the blades (Figure 2.3). An unshouldered paddle has a smooth transition from the paddle-shaft to the blades, and it's difficult to tell where the shaft stops and the roots of the blades begin. Both styles continue to be used in Greenland.

An unshouldered paddle has the advantage that it does not dictate hand position, as does a shouldered paddle. This may make paddle fit slightly less critical and permit kayakers to purchase acceptable,

noncustomized commercial paddles. Some kayakers prefer unshouldered paddles, claiming that the smooth blade transition is less likely to impede their sliding strokes, but I don't find this to be an issue with either paddle type, if properly shaped. Unshouldered paddles are also easier to make.

For shouldered paddles, the paddle-shaft length and the shoulder shape are critical to enjoyment and maximum performance. If properly made, shouldered paddles allow the kayaker to center the paddle easily between his or her hands by feel. This is a useful feature, since most kayakers' hands roam the paddle frequently. However, if the shoulders are too close together, temporary wrist pain can occur due to the side-to-side wrist flexion that results (unless a sliding stroke is used). Conversely, if the shoulders are spread too far apart, all of the kayaker's fingers will fall on the shaft. This produces much less awareness of blade orientation than actually grasping the blades and does not encourage the blades to tilt forward for the canted blade stroke.

For a shouldered paddle that is normally used without a sliding stroke, I currently prefer a paddle-shaft length of 20 inches (about 50 cm).

There is tremendous variation in paddle-shaft length in Greenland. Petersen writes, "The [paddle-shaft] can [be found in] different lengths. In Kangaamiut David Rosing measured a [paddle-shaft] only 25 cm (9 ¾ inches) long while those of longer paddles are longer than the width of the kayak" (Petersen 1986:66).

A simple method to determine the paddle-shaft length for a shouldered paddle is to stand naturally and allow your arms to dangle freely (not pressed against your sides). Raise your forearms by bending

Figure 2.3—Detail of shouldered Greenland paddle. Shoulder profiles can be very abrupt or quite subtle—the only requirement is that they comfortably fit your hands. Photo by Greg Stamer.

only at the elbows so that your forearms are horizontal and parallel to each other. If you make a circle with the thumb and forefinger of each hand, this marks each paddle shoulder (where the paddle-shaft should transition into the blades). This method is based upon my own experimentation and is not a method that I have been shown in Greenland. However, my paddles compared very favorably with those that I saw and used at the training camp I attended in Qoornoq. Alternately, some kayakers prefer experimenting with paddle-shaft length by holding a large dowel between their hands (a discarded broomstick works) while seated in the cockpit and simulating paddling strokes, to determine where their hands naturally fall during use. An advantage of this method is that it takes into account the unique attributes of both the kayaker's body and the kayak. A kayaker may want to make his or her first paddle with the paddle-shaft slightly shorter than necessary to allow for fine-tuning over time.

Some Greenlanders prefer a shorter paddle-shaft length of approximately body width, as opposed to the shoulder-width measurement described above. According to Maligiaq Padilla, this sizing is often intended for use with a partial sliding stroke where the paddle is shifted in the hands only several inches on each stroke. Ivars Silis's 1995 video, *Amphibious Man*, has excellent footage of John Petersen demonstrating this technique.

Compared to those of his peers, Maligiaq Padilla's paddle-shafts are relatively wide at 24 inches (61 cm), although this is still short compared to a contemporary sports paddle. In Padilla's case, this dimension is not based upon anthropometric measurements. He copied this measurement directly from the paddle of one of his kayaking coaches and found that it worked well for him. According to Padilla, the slightly wider hand position is more advantageous for racing.

In stormy locations, shorter paddles with extremely short paddle-shafts were developed that are used exclusively with a sliding stroke. Although the paddle-shaft length can vary, Kâlêraq Bech (Sisimiut) prefers a measurement of only three fists wide. This short paddle, often called a storm paddle outside of Greenland, has much to recommend it as a spare paddle, since it takes up relatively little room on deck and, unlike a two-piece paddle, doesn't have to be assembled in an emergency prior to use. Despite the current popularity of spares, they were not traditionally carried in Greenland (Petersen 1986:67–68). I have seen spare paddles used during races at the Greenland championships, but the practice is fairly rare.

The circumference of Greenland paddle-shafts is sometimes larger than that of sports paddles; a rough measurement is the circle formed by pressing the tips of your thumb and forefinger together. Whereas sports paddles are often rounded or oval in cross section, Greenland paddle-shafts—although there are many variations—can be almost rectangular in cross section with rounded corners, the long axis of the paddle-shaft being perpendicular to the blades for strength and ergonomics. This shape makes it quite easy to determine the orientation of the paddle blades through feel alone.

BLADE WIDTH AND SHAPE

Petersen describes the blade width of a Greenland paddle as "the distance from the innermost joint of the thumb to the middle joint of the other fingers" (Petersen 1986: 65); however, this dimension can vary widely. The paddle must be narrow enough to permit a secure grip, but different kayakers will prefer different dimensions. For example, Maligiaq Padilla, who is thin and strong, prefers a paddle blade that is only 2 5/8 inches (6.7 cm) wide. Padilla has experimented with wider blades but reports that doing so only decreased his speeds. By contrast, Steffen Olsen, strong and stocky and one of the top kayakers in the older men's championship class, prefers a paddle almost as wide as he can grip, about 5 inches (12.7 cm), for increased resistance.

The edges of a Greenland paddle must be well formed and properly maintained to provide comfort when gripped in the kayaker's hands. With this in mind, I was surprised at the "sharpness" of the edges of several Greenland competition "racing" paddles that I examined at a training camp in Qoornoq in 2000. Unlike the blunt edges that I was accustomed to seeing and using, these paddles, although they retained thickness in the center of the blades, had well-defined edges. Likewise, the paddle tips were thin and sharp-edged. These paddles subjectively demonstrated a greater "bite" in the water and deserve more study. The lack of bone edging, combined with the sharp edges, makes them less durable. Perhaps these paddles are a recent variation for racing, as hunting paddles are often made with rounded edges for stealth (Petersen 1986:65).

PADDLING STROKES

Canted Blade Forward Stroke. Maligiaq Padilla commonly teaches two main paddling strokes, a beginner's stroke with the paddle blade oriented vertically to the water's surface and an advanced stroke with the paddle blade canted slightly forward, so that the top edge of the paddle blade inclines closer to the bow of the kayak than the bottom edge (Heath 2000:56–57). (Note that for both techniques, the described blade orientation refers to rotation about the long axis of the paddle.) The canted blade stroke appears to be a fairly common technique in Greenland, and I have discussed it with kayakers residing from Nuuk to Upernavik. Although the regional and temporal extent of this technique is currently unknown, knowledge of this stroke is not unique to Greenland. Eugene Arima provides the following description of this technique once used with the Caribou Inuit kayak in the central Canadian Arctic:

> Paddling was done with long but only moderately deep strokes because if the paddle was dug too deeply in the water, one could follow it downward. Grip was taken with the hands placed over the shaft as far apart as was comfortable with a wide range of individual variation from about 20 inches [about 50 cm] apart … to only 8 or 9 inches [20 to 23 cm] between the hands…. The blades were held at a slight angle from the perpendicular, upper edges forward of the lower ones, to be changed readily to the position for the planing paddle brace by increasing the angle. To counteract a sideways lean, the kayaker quickly lowered the paddle near the deck by extending the arms forward and downward, simultaneously turning the hands down from the wrists to place the blades at a suitable angle for planing on the water surface, leading edges slightly higher than the trailing ones. The paddle hardly had to be dipped on the side of the lean to bring the blade against the water. This planing brace was also used to turn or stop the kayak. The adept could take a drink of water with its aid (Arima 1975:145–146).

When adapting this information to Greenland-style kayaking, note that the Caribou Inuit kayak was extremely fast and unstable, often used for hunting swimming caribou, and that the kayak and paddles used were quite different from those typically used in Greenland. Such a kayak could be more than 26 feet long (about 8 m) with a width of about 18 inches (45 cm) and was used with a narrow paddle at least 13 feet (about 4 m) long.

The canted blade technique remained somewhat controversial outside of Greenland, at least in the United States, until Maligiaq Padilla toured the United States and Canada in 1998. For almost a year, he performed numerous kayak demonstrations popularizing the technique.

Canting the blades creates a number of benefits, including controlling flutter and ventilation, enabling the paddle blades to bury quickly, as well as providing a stronger "bite." Initially, however, kayakers may find that this technique feels somewhat perilous due to the sensation that the blades, which dive and bury quickly during the stroke, will cause a capsize. Generally this sensation disappears after a short period of practice.

Some novices initially feel that a Greenland paddle "slips" too much, resulting in slower forward speeds. This can lead to some worry about being able to keep up with the pack. Learning to cant the blade can readily improve the feeling of resistance, lower paddling cadence, and produce higher speeds. Chris Cunningham, in an analysis of Maligiaq Padilla's stroke, writes: "Switching from the beginner's stroke to the [canted blade] stroke was very much like switching between a standard paddle and a wing paddle. The increase in the pull on the paddle is readily apparent" (Cunningham 2000:56).

Although "feel" can sometimes fool you, the feel of water flowing over the blade during the canted blade stroke is very different from the sensation of simply dragging a vertical blade straight backward. This has given rise to theories that lift is more predominant with the canted blade stroke. Unfortunately, there have been no definitive studies to date, and there is still much debate over the physics behind an efficient paddle stroke and the role of lift versus drag. Jeffrey Wilkes provides an excellent synopsis:

> There are basically two approaches [to paddling] one can follow. If [a] paddle is placed in the water, held vertically, and then pulled backwards with a large force then a relatively small amount of water will be imparted with a large velocity. Two vortices (rotation in opposite directions) will be seen spiraling off the paddle. This approach is common among novice canoers….
>
> A better method is to put the paddle in the water on the fly with an angle of attack such that lift is produced. In this case only one vortex is seen spinning off; the other vortex stays with the blade and affects a large mass of water (Wilkes 1999).

In the case of a canted blade stroke, it is fairly easy to demonstrate the latter effect—one vortex being shed after the catch, with the other vortex bound to the blade and shed only upon the exit of the stroke.

Basic Mechanics of the Stroke. The canted blade stroke is not difficult to perform, although it requires some time to master. Since it is difficult to instill a mental image of this stroke by focusing on fine details, following is a simple drill to help the student understand the overall body mechanics. Note that in an actual stroke, power comes not from the kayaker's arms but from the large muscles of the legs, back, and abdomen.

Stand casually, without holding a paddle, and with arms bent so that your forearms are horizontal and parallel to each other. Your elbows will point downward but your arms should not be pressed against your sides. To get a feeling for the arm movement, simply raise your right forearm into a near vertical position by bending at the elbow and optionally allowing your elbow to "fly" (move away from your torso) to a small degree. This will place your hand between shoulder and eye level. Lower this hand to its original position by levering your forearm forward while keeping your elbow pointed downward and bent. Your hand will move straight ahead while moving downward. This motion is then repeated on your other side. Generally your elbows remain at least partially bent throughout the stroke. Once you understand this basic arm motion, add torso rotation. As your right hand moves forward (and downward), rotate your torso counter-clockwise (and vice versa). Notice how this will cause your "pushing" hand to follow your torso rotation. Due to this torso rotation and the relatively close spread of your hands, your pushing hand will regularly cross over the centerline of your kayak deck rather than traveling straight along the gunwales.

Although this drill is vastly oversimplified, it helps to demonstrate the economical range of motion for the stroke. This economy of motion requires less energy expenditure, as the kayaker is required to lift his or her arms much less than in some sports kayaking techniques. Instead, the kayaker's hands stay fairly low during the stroke. Unlike wing paddle technique, where the elbows are raised into a high "chicken-wing" position, the kayaker's elbows will fly somewhat but will remain pointed downward.

Forward Stroke, Holding the Paddle. A wealth of technical details are visible in the photograph of Kâlêraq Bech (Figure 2.4). Note how his elbows are close to his sides, but with enough clearance for freedom of motion. Although his pushing hand is relatively low, having started at about shoulder level, note that the paddle tips can still be quite elevated. This is due to the long blades and his relatively close hand separation. Also examine his grip. Bech's thumbs and forefingers are on the paddle-shaft; his other fingers are draped over the blade roots. This causes the paddle to tilt forward slightly, evident from the paddle tip. Gronseth (1992:51) writes:

For me, learning to kayak with the narrow-bladed Greenland paddle was like learning to kayak all over again. Even the way the paddle is held for a forward stroke is different. Unlike the way our modern paddles are used, a Greenland paddle is held with the top edge tipped slightly forward. The last three fingers of each hand grip the beginning of the blades.

Initially, the idea of holding a kayak paddle with the blades tilted forward might sound like a very contrived and perhaps precarious way to paddle. However, with a well-fitting paddle, the forward tilt of the blades will occur without much conscious thought. Holding the paddle with only the thumbs and forefingers around the shaft, with the remaining fingers draped over the roots of the blades, allows for a delicate yet secure grip and excellent feel for the orientation of the blades. This causes the paddle blades to tilt forward slightly because the palm of the kayaker's opened hand is also angled forward when his or her wrists are held in a neutral (uncocked) position. When holding a paddle in this manner, the kayaker would actually

Figure 2.4—Kâlêraq Bech at 2000 Qaannat Kattuffiat training camp in Qoornoq, Greenland. Photo by Greg Stamer.

have to cock his or her wrist slightly backward in order to obtain a perpendicular blade position. Although there has been much confusion about the "proper" blade angle to use, the forward tilting of the paddle should occur naturally and is subject to personal variation.

If the paddle is sized so that all of the kayaker's fingers fall on the shaft, the paddle blades will often tend to orient vertically to the water's surface. Even though this may make the canted blade technique counter-intuitive, experimentation is still possible by manually adjusting how the paddle sits in the kayaker's hands. Recently, as Greenland kayaking and mainstream kayaking techniques have started to cross-pollinate, some kayakers using contemporary sports paddles have also reported benefits to canting their blades.

Forward Stroke, Stroke Foundation. A strong paddle stroke is difficult to produce unless the kayaker's lower body is solidly grounded. Instruction for mainstream kayaking emphasizes the use of the leg muscles and torso rotation over the use of the comparatively weak arms alone; the same philosophy is found in Greenland among the kayakers I have interviewed. However, when I asked a small number of Greenlanders participating in the annual competition to elaborate upon how they used their feet and legs, I received conflicting answers. Maligiaq Padilla gave me one answer, and then after paddling for an afternoon and reflecting upon his technique, changed it. The consensus is that solid lower body foundation is important, but different paddlers achieve it in different ways, either by foot/leg resistance or by knee/leg resistance or both.

- Foot/leg resistance is taught in mainstream kayaking instruction and is also used by some Greenlanders. For a stroke on the paddler's right side, the stroke starts by first relaxing the left foot on the foot braces and then pressing the right foot and leg into the foot braces. This allows the paddler's entire right side to act as a foundation, providing resistance and helping to drive the strong torso rotation that follows. The feet "dance" on the foot braces: one foot/leg goes forward while the other relaxes, alternating on each stroke. Alternating foot pressure has the added benefit of helping to prevent the paddler's feet from falling asleep.
- Knee/leg resistance is bracing with the knee opposite to the side of the paddle stroke. This is a very natural technique in a true Greenland kayak due to the low *masik* (curved deck beam that acts as a thigh brace) that requires a fairly straight-legged paddling position. As the foot/leg on the side of the working paddle blade moves forward to engage the foot brace, the opposite knee

will rise slightly. For a stroke on the right side, as the right leg presses forward into the foot brace, the paddler's left knee will rise and strongly engage the *masik*. While this effect is rarely noticed in high-volume kayaks, kayaks with a very low foredeck, such as a Greenland kayak, will often generate the sensation that the resistance of the opposite knee as it engages the *masik* is much more dominant than foot pressure on the side of the stroke. High-volume kayaks may require padding added to the underside of the foredeck to permit this technique. Finally, this technique is often complemented by a strong contraction of the abdominal muscles to help drive the paddle using large muscle groups.
- Depending on the height of the foredeck, both of the above methods can sometimes be employed.

Forward Stroke, Catch Phase. The catch phase, the initial contact of the paddle blade with the water, should be initiated by rotating the paddler's torso, but unlike the modern racing stroke, there is no attempt to reach as far forward as possible. Kâlêraq Bech teaches that the torso should be relaxed. For someone trained in sport technique, the catch with a Greenland paddle will feel much farther aft. The canted blade will help to ensure a clean catch, that is, smooth insertion of the blade without ventilation or flutter.

Ventilation occurs when air is sucked underwater with the blade, resulting in a loss of efficiency. Fortunately, ventilation is very easy to detect with a Greenland paddle as it is announced by a scratching sound, very similar to drawing your fingernails over coarse nylon fabric. To prevent ventilation, the catch can be initiated with a quick downward movement of the angled blade, where the blade slices edge first into the water. Another technique that works very well is to adopt the wing paddle technique commonly called "spearing the salmon." This is performed by allowing the upper hand to initiate the catch with a lateral thrust of the paddle tip into the water. Ventilation can also be greatly reduced by ensuring that the working blade is fully buried up to the paddler's lower (pulling) hand before applying full power. The diving blade angle will help to ensure that the working paddle blade buries very quickly and completely.

Flutter, an uncontrolled zigzagging of the paddle as it moves through the water, is commonly caused by pulling straight back on a blade oriented at a right angle to the water. Doing so causes vortices to break off alternate edges of the blade; the different lift vectors that result cause the blade to move in different directions. By canting the blade, the vortices break consistently along a single edge and can be controlled, eliminating flutter.

A common error is to plant the paddle too far forward and over-reach. This can cause the paddler's wrist to cock backward, transforming the blade angle from top edge tilted forward to the lower edge tilted forward. This makes it very difficult for the paddle to enter the water cleanly; ventilation and a noisy "plop" often result.

When touring I have noticed that many Greenlanders use an easy, gentle catch with power building through to the exit. Doing so will lessen the initial shock to the kayaker's musculoskeletal system and may help to prevent injuries.

Forward Stroke, Torso Rotation, and the Abdominal Crunch. When first exposed to an expert kayaker from Greenland, many kayakers trained in mainstream technique commonly exclaim that the Greenland stroke is devoid of torso rotation. Actually, Greenland kayakers use strong torso rotation, but it is visually subtler than the extreme side-to-side rotation used by kayakers experienced with wing technique. In addition to side-to-side torso rotation, many Greenlanders also move diagonally forward and downward, similar to performing a sit-up directed toward the opposite knee. The torso starts erect and then lowers slightly, bending from the hips with a straight spine, powered by the abdominal muscles working together with strong knee pressure against the *masik*. The knee opposite the working paddle blade provides the foundation and the effect is to use very large muscle groups to help drive the paddle. The abdominal crunch is visually very subtle at touring speeds, imparting a smooth rhythm, but is very pronounced for sprinting. Proper timing is critical, and the "crunch" takes place only after the working paddle blade is fully buried. Kâlêraq Bech teaches that the abdominal crunch, if used at all, should be smooth, and that the kayak must not bounce as a result.

An additional benefit of using the abdominal crunch is that it relieves pressure against the backrest, making an elaborate padded backrest unnecessary (most Greenlanders simply use a thin foam pad that extends from the aft cockpit rim to the footrest). The abdominal crunch also works well even when you are required to tuck low toward the foredeck to reduce your wind profile, a position that can greatly inhibit normal torso rotation.

Forward Stroke, Use of the Arms. Independent use of the arms, or "arm-paddling," is discouraged by many mainstream kayaking instructors in order to ensure that the larger muscle groups dominate. Proponents of arm-paddling believe that the use of the arms, if linked to the rotation of the torso, provides additional power. In doing so, the upper hand "pushes" forward in a slow-motion "punch" and almost straightens while the lower hand "pulls," creating a fulcrum centered about the sternum. Most Greenlanders that I have observed keep their arms fairly bent throughout their stroke, which indicates that torso rotation predominates. Regardless of which technique is used, the paddler will, of course, feel pressure against the hands. Maligiaq Padilla reports feeling about 60 percent push and 40 percent pull during his stroke. He often opens his pushing hand to reduce tension and promote circulation—good to prevent the hands from becoming cold or numb—and his pulling hand often draws the paddle backward with his fingers forming a loose hook. Those who experiment with active arms should realize that unless the kayaker is using a sliding stroke, which permits a wider hand separation, the effect is subtler than similar techniques using sports paddles, since the hands tend to be much closer together with a Greenland paddle.

Forward Stroke, Power Phase. I find that the paddle is at its deepest point just after the catch, with the paddle blade buried almost to my pulling hand, and that the paddle rises slowly during the remainder of the stroke, to the exit. I find this down/up movement to be fairly subtle, but George Gronseth provides a slightly different description based upon his Greenland training camp experience: "With a Greenland paddle, the motions of the forward stroke are more down/up than front/back.... [This] allows the working blade to generate lift in the forward direction, which the kayaker turns into forward propulsion" (Gronseth 1992:55). Maligiaq Padilla, when asked about this technique, responded that he is not familiar with the use of a strong down/up movement for a forward stroke, but he speculated that this could be a local variation.

The path of the paddle, when viewed from above, is also subject to variation. Some kayakers, such as Maligiaq Padilla, allow the working paddle blade to move almost straight back (parallel to the keel). Alternatively, the paddle can be allowed to flare out away from the gunwale, somewhat reminiscent of a wing-paddle stroke. This works quite well with a canted blade stroke. According to Kâlêraq Bech, there is also a specialized variation where the wrists flick the blade toward the stern at high speeds; however, this method appears to be uncommon.

Forward Stroke, Exit Phase. Mainstream instruction often teaches that the working paddle blade must be withdrawn from the water before it reaches the paddler's hip, to avoid lifting water. However, Greenland blades, being much longer, will naturally pass behind the hip on all but the shortest paddle strokes. In the case of the forward-canted blade, although water is lifted as the angled blade moves upward during the latter part of the stroke, additional forward thrust is also generated. The mechanics are similar to sculling and can be easily demonstrated by the following drill. If you rotate your torso and immerse the blade of a Greenland paddle slightly behind your hip, cant the blade forward, lift it straight upward, and then repeat on the other side, your kayak will slowly move forward. This produces additional thrust that you can tap for your stroke. In actual use, you would not want to lift your hands upward as in this drill. Your "pushing" hand moving forward and downward causes the upward movement of the working blade during an actual stroke.

Maligiaq Padilla uses slightly different mechanics for the exit of his stroke. Rather than allowing the paddle to lift significantly upward toward the end of the stroke, he allows the paddle to lift slightly upward before exiting the water with the blade slicing out edge first, moving toward the bow. This technique is valued in heavy weather, as the blade is essentially feathered as it rises out of the water, making it less likely to be slapped or stopped by waves (Heath 2000:56).

Forward Stroke, Sliding Stroke. A sliding stroke limits the amount of blade area that the airborne blade exposes to the wind, reducing wind resistance. A full sliding stroke used with a short storm paddle is often complemented by the use of the canted blade technique. When paddling into strong head winds with a storm paddle, the paddle shaft is often held vertically, the blades passing close to the hull of the kayak. The kayaker's torso is often tucked forward toward the deck—similar to being on "the drops" of a racing bicycle—in order to reduce wind profile. Note that although a strong forward tuck can impede torso rotation, it does not impede the use of the abdominal crunch technique, which can prove very useful in this situation. A full sliding stroke can also be used with a full-sized paddle. With the paddle-shaft held nearly vertically, this stroke is useful for performing intense, short-duration sprints. With the paddle-shaft held horizontally, a sliding stroke can allow navigation of extremely shallow water. In addition to limiting blade exposure to wind, this stroke can prove useful in other situations. I occasionally use a full sliding stroke in mangrove tunnels and other areas of low-hanging vegetation in order to prevent the outboard blade from catching the foliage.

To perform a full sliding stroke, start with your left hand grasping the center of the paddle-shaft, with your right hand securely grasping the blade near the tip—not cupped around the tip. Your hands will be approximately shoulder-width apart. Take a stroke on the left side. As the working paddle blade exits the water, slide your right hand—the hand closest to the paddle tip—down the paddle-shaft until both hands meet or almost meet. Then, as your torso rotates into position to allow for a stroke on your right side, allow your torso rotation and a loose grip of your left hand to pull the paddle through that hand; regrip near the paddle tip to start the next cycle. Although initially the motion will seem quite "busy," much of the paddle movement is driven by your torso rather than by independent arm action.

A partial-sliding stroke, used with a full-size paddle, is another useful option. This stroke is similar to the full-sliding stroke, except that the kayaker shifts the paddle only a few inches or more per stroke. This technique allows the kayaker to vary the amount of blade resistance in the water. It is also very useful when using a paddle that has a short paddle-shaft, to allow comfortable hand separation. Some Greenlanders use this technique as their primary forward stroke.

Forward Stroke, Paddle Height. One of my initial surprises in studying Greenland technique at the 2000 training camp and from working with Maligiaq Padilla was learning that holding the paddle low and horizontally was not a fundamental technique in Greenland. Although a low horizontal stroke appears to have been common in some parts of the Arctic, particularly where very long paddles were used—and in some cases with the paddle-shafts resting on the foredeck—in Greenland the height of the paddle varies depending on how much power the kayaker needs to exert. A very low stroke can be useful for economical cruising and for rough weather where every stroke acts as a brace; but for touring, many kayakers that I observed held the paddle at approximately 45 degrees to the horizon. During sprints, the paddle is held as vertically as possible with the working blade passing close to the gunwale for maximum thrust. I tend to view paddle height with a Greenland paddle similar to the gears on a bicycle: I vary the height depending on the weather conditions and how much speed I wish to achieve. Do not confuse paddle height with the height of your hands. Since your hands are spaced relatively close together and the blades are long, you can sprint while holding the paddle-shaft almost vertical while still keeping your hands at or below eye level.

PADDLE EXTENSION For turning the kayak using sweep strokes or rudder strokes and for bracing and rolling, a Greenland paddle is designed to be extended. This allows even a short paddle to generate maximum leverage and lift. This practice is a departure from some whitewater and contemporary sea kayak instruction that often teaches a fixed hand position. One argument in favor of a fixed hand position is that shifting the kayaker's hands is an invitation to lose the paddle in violent weather. Another common argument is that in performing a lightning-fast reflexive brace, the kayaker simply does not have time to extend the paddle. Both arguments, when directed toward Greenland-style kayakers, are based upon an incorrect understanding of Greenland kayaking equipment and technique.

The key to safely extending the paddle is to ensure that one hand always retains a secure grip while the other hand slides, and that this extension is accomplished with one, or at most two, movements. After paddle extension and prior to bracing, both hands should have a secure grip on the paddle. As an example of a quick reflexive brace, imagine the kayaker seated in a kayak; a wave upsets his or her balance so that the kayaker is falling toward the right. The right hand should retain its secure grip on the paddle (near the paddle shoulder); as the kayaker's torso leans toward the right to perform the brace, he or she temporarily relaxes the left hand so that the paddle can slide through it. The result is an extended paddle on the side of the brace. Note that the actual extension is accomplished by the kayaker's torso leaning toward the side of the brace, rather than by independent hand and arm movements. Paddle extension of this type can be done extremely quickly once proficient.

A skilled and attentive kayaker often has ample time to anticipate a brace. In this situation, he or she can achieve full paddle extension by using a two-step method. Again, using a brace on the right side as an example, the kayaker first slides the right hand a short distance leftward, to securely grip the paddle-shaft near its center. Then, as the kayaker leans right into the brace, he or she temporarily relaxes the grip with the left hand, allows the movement of the paddle to slide through it, and grips the paddle blade several inches from the tip.

When fully extending the paddle, it is not customary in Greenland to "cup" the palm around the extreme end of the paddle tip. While cupping is sometimes permitted under the rules for a few very contorted competition rolls at the Greenland National Kayaking Championship, this practice is strongly discouraged according to my Greenland informants. Cupping the paddle tip requires the kayaker to remove or partially remove his or her hand from the paddle in order to rotate the hand into position, making this a potentially risky movement. Cupping is often used in lieu of appropriate torso rotation for rolling and sculling, thus moving the torso and arms out of a stronger biomechanical position. Finally, this practice may break mortised paddle tips, which are very susceptible to strain.

BRACING TECHNIQUE   A kayak provides a certain measure of static stability, but the seaworthiness of a kayak is largely provided by dynamic stability in the form of braces and technique applied by the kayaker. Low and high braces, similar to what is taught for mainstream kayaking, are also used in Greenland. A low brace is performed with the paddle held below the kayaker's wrists and elbows, with palms facing downward. The back of the blade—the side of the blade that faces away from the kayaker during a forward stroke—is presented to the water. A high brace is performed with the paddle-shaft held below neck level, palms facing upward, and the "powerface"—the side of the blade that faces the kayaker during a forward stroke—presented to the water. To avoid shoulder injury, the paddler should rotate his or her torso to keep the paddle-shaft facing the chest as much as possible. A basic brace involves getting purchase or lift with the paddle, and then rotating the kayak back under the body using a hip snap, or rotation of the lower body. Full or partial paddle extension is commonly used with a Greenland paddle during a brace; some practitioners add a forward or reverse sweep during a brace to generate lift and to aid recovery.

The recovery phase is performed differently with Greenland paddles than is often taught by mainstream instructors. Rather than recovering by rotating the paddle blade on edge and slicing it edgewise up and out of the water, the working blade of a Greenland paddle is kept pressed flat against the water (Figure 2.5). To recover, the entire paddle is slid as a unit over the foredeck—the kayaker's hands hold the paddle securely and do not slide over the paddle during this movement. Since the paddle is held inclined with the inboard blade higher than the immersed working blade, moving the paddle longitudinally over the foredeck, with the shaft moving perpendicular to the keel, generates lift. This also keeps the blade properly oriented and immersed, should the paddler need to add a sweep or sculling action to assist with recovery. To picture this recovery method for a low brace, imagine a fisherman leaning over a low gunwale and pulling a net into his boat.

Figure 2.5—Low brace recovery. The working blade is kept pressed flat against the water and the paddle is slid as a unit over the foredeck. Photo courtesy Gary Landwirth.

SIDE SCULLING

Greenland kayakers are often first taught to remain calm while capsized, and then how to scull before moving on to rolling. Sculling is a technique that depends upon finding a balanced position while the paddle is swept back and forth to produce continuous lift. Learning to scull before rolling is a very logical progression; once kayakers can place their ear in the water and scull up, they will find that they can use the same technique to recover from a capsize. Learning this skill opens the door to several new techniques, including the balance, or static, brace. The balance brace is a related technique that allows the paddler to find a balance point requiring no paddle motion at all. One benefit of performing both the balance brace and side sculling is that these techniques enable the paddler to fully stretch his or her lower back and legs. This can prove very useful while constrained for long hours in a cramped cockpit.

An excellent way to learn a side-scull is to have an instructor help manipulate the kayak and the kayaker's body to discover the balance position. To learn to side-scull alone, place a float on the extended paddle blade. This technique is performed in a high brace position, meaning that the palms face upward; the paddle will be held just under the chin, with elbows bent and fairly close to the paddler's sides. The following instructions are to perform a side-scull on the right side: Fully extend the paddle to the right and strongly rotate your torso counterclockwise so as to present your back as

flat to the water as possible. If you are not very flexible, this can be more easily accomplished by shifting in the cockpit so that you are sitting somewhat on the side of the seat. The working blade will be immersed about 45 degrees off the bow of the kayak, rather than at right angles to the keel as is commonly assumed, because of strong torso rotation and because the paddle-shaft follows the line of the kayaker's shoulders.

The action of the lower body is important to understand. When the kayaker performs a high or low brace with the torso held above water, he or she engages the knee opposite the working blade for support. For example, when leaning toward the right, the left knee will press hard against the *masik* or the thigh braces. However, once the kayaker's torso becomes immersed, as in a side-scull, a transition occurs. The pressure changes from against the upper knee/thigh to that of the lower leg, which should be fully extended. If the kayak has enough volume, the kayaker can even drop and lay the upper leg upon the lower leg to get his or her weight as close to the water as possible. I like to think of my weight as molten lead and flow as close to the surface of the water as possible. In this regard, contemporary commercial kayaks and outfitting have a disadvantage. Typically, a Greenland kayak is custom-built with just enough clearance for the hips to allow the kayaker to rotate in the cockpit. Commercial kayaks often require the kayaker to add hip pads in order to create a secure internal fit. When the kayak is heeled over on its side, these pads lift the lower body and prevent the kayaker from getting his or her weight as low as possible.

Once in position, fall backwards. As soon as your back touches the water and you change your foot pressure to the lower leg, you need to push the kayak away from you. If the kayak deck sits at a right or greater angle to the surface, the kayak will simply fall over on you, pushing your torso under the water. You need to counteract this tendency and keep the kayak on as even a keel as possible, at about a 45-degree angle. How you accomplish this will depend upon your fit in your kayak, but the most common technique is to strongly arch your back. Some kayakers find that they have to push with their legs as well. With practice, you will find a balance point where you can sit motionless, facing straight up with your back nearly flat on the water and your head immersed for buoyancy.

From this position, you can add a slow, smooth sculling motion to generate lift (Figure 2.6). You must keep the leading edge of the blade elevated during the sweep, which means that you will alter the blade angle every time you reach the end of a sweep and have to change direction—this motion is similar to using a knife to spread icing on a cake. A Greenland paddle is very lenient of sculling technique and only a subtle change in blade angle is required. All control of the sculling motion is accomplished using your lower, or outboard, hand. Your upper hand can be opened if you wish; it simply provides a support platform for the paddle. The buoyancy of the Greenland paddle aids in keeping your torso on the surface.

To recover to an upright position, perform a sweep to generate lift and add a hip snap to roll the kayak back under you. With practice, this motion can be very subtle and a proficient kayaker will simply appear to sit upright. For example, to recover on your right side, perform a strong sweep on the right and then relax your left leg/foot before briskly lifting your right knee into the *masik* to generate rotary motion.

The side-scull is a very useful technique for capsize prevention and recovery. Maligiaq Padilla taught me that if the kayak should be violently thrown on its side, rather than bracing forcefully in an attempt to keep the body above water, and potentially risking shoulder injury, the kayaker should adopt the side-sculling position instead. Hitting the water nearly flat on your back while pushing the kayak away immediately arrests a potential capsize, using the

water as an ally rather than fighting against it. Once in a side-sculling position, you can scull for as long as is necessary and then recover to an upright position at your whim.

In addition to low and high braces, Greenlanders perform a technique in which the kayaker braces from side to side with his chest facing to the side of the kayak, a maneuver that I call a "chest brace." John Petersen demonstrates this in John Heath's *Greenlanders at Kodiak* videotape (Heath 1989). To perform a chest brace, the kayaker's torso faces one side and the paddle is extended to that side, perpendicular to the keel (Figure 2.7). Some Greenlanders prefer to rotate their outboard hand so that their thumb points toward the paddle tip. This technique requires commitment, as the kayaker must lean over and immerse his or her inboard hand so that the paddle is held at an angle, with the outboard end elevated above the water's surface. To recover, the kayaker performs a forward-leaning hip snap while at the same time quickly lowering the paddle to a horizontal position. A closely related technique, the chest scull, uses the same body position and relies on a sculling paddle blade for continuous support. You can hold your head above the water or keep it immersed. You drive the sculling action by torso rotation and fairly passive arms. Recovery is the same as for the chest brace.

Figure 2.6—Side sculling. The kayaker's torso and head are immersed for buoyancy. This technique relies on finding a balance position and from lift produced by the sculling blade. Control of the sculling motion is accomplished with the lower hand. The upper hand can be opened and acts merely as a support platform. Photo courtesy Gary Landwirth.

Figure 2.7—Chest brace. The kayaker's torso faces the side of the brace. The elevated paddle is quickly lowered at the time of the hipsnap to recover. Some Greenlanders practice very quick braces using this technique, alternating sides on each repetition. Photo courtesy Gary Landwirth.

PADDLE MAKING

Commercial sports paddles and fiberglass kayaks are available in Greenland, but I have seen this equipment used only for the tourist industry, with few exceptions. Economics, tradition, and the excellent performance of Greenland kayaks and paddles are probably key factors. Furthermore, the vast majority of Greenlanders build their own equipment.

Making a Greenland paddle saves money and results in a custom-sizing. With a quality contemporary paddle ranging from $150 to over $450, making your own paddle from reasonably priced lumber, with a few simple tools, has tremendous appeal. I use a drawknife, a high-carbon steel woodworking knife (such as a sloyd knife), a block plane, and a spokeshave for the majority of the work, but power tools such as a band saw can be real time-savers. Some modern Greenlanders make extensive use of power tools for paddle and kayak construction.

Traditionally, huge logs of driftwood were split into manageable sizes for kayak and paddle building, although today most Greenlanders purchase their lumber. Many types of softwood are satisfactory for paddle building. In the United States and Canada, western red cedar, Douglas fir, spruce, and pine are commonly used.

Careful selection of wood is essential. Clear (knot-free), vertical grain (quarter-sawn) lumber is an excellent choice for strength, stiffness, stability, and resistance to warping. Vertical grain lumber can be identified by the grain—annular growth rings—that appears as tight pinstripes running along the face of the board from end to end. If viewed from the ends, the grain runs straight up and down. Flat-sawn lumber—end-grain parallel to the width of the board—is also used in Greenland, but among the competition paddles that I have inspected, it is not as common a choice as quarter-sawn. Flat-sawn lumber tends to be cheaper and easier to find, but for a solid, non-laminated paddle, I find it too flexible. This is especially true when the paddle is used to perform advanced sculling rolls, for example, with the paddle held on the foredeck or under the hull; the paddle will lose lift or even threaten to break if it bows too much. Runout, where the grain runs diagonally off the side of the board, should be avoided, especially if it occurs in the paddle-shaft area. Grain runout is a common cause of paddle failure.

Greenland paddles, especially those used for hunting and competition, must be extremely strong. As a case in point, one of the techniques performed at the annual Greenland National Kayaking Championship is the "walrus-pull." This involves five men on shore pulling a kayaker sideways via a line attached to the kayak, while the kayaker fights to prevent a capsize. This pull places tremendous strain on paddles, and injuries have occurred from paddle breakage. I competed with a lightweight, western red cedar paddle in 2000. My fellow competitors complimented this paddle when they hefted it in their hands. Although it survived my walrus pull attempts, competition judge Kamp Absalosen voiced his disapproval. In Absalosen's opinion, the paddle was not robust enough. He stated that a paddle must be very strong—strong enough to be used as a chin-up bar.

References for building a Greenland paddle may be found on the Internet or purchased from several sources. Paddle and kayak building references and video clips of Greenlanders performing rolls, braces, and forward stroke variations are available on the website of Qajaq USA (www.qajaqusa.org), the American chapter of the Greenland Kayaking Association.

REFERENCES

Arima, Eugene Y.

1975 *A Contextual Study of the Caribou Eskimo Kayak*. Ottawa: National Museum of Man, Canadian Ethnology Service, Mercury Series no. 25.

Cunningham, Christopher

2000 Maligiaq's Forward Stroke: An Analysis. *Sea Kayaker* 17(2):56.

Golden, Harvey

2003 Qajaq USA Greenland Forum. Retrieved May 11, 2003, from http://www.qajaqusa.org/cgi-bin/GreenlandTechniqueForum_config.pl/noframes/read/10884.

Gronseth, George

1992 Learning to Kayak the Greenland Way. *Sea Kayaker* 8(4): 46–55.

Heath, John D.

1986 The Narrow Blade—Theory and Practice. *Sea Kayaker* 3(1): 13 16.

1987 The Do-It-Yourselfer's Greenland Paddle. *Sea Kayaker* 4(3): 58–59.

1989 *Greenlanders at Kodiak*. Video produced and directed by John D. Heath; 38 minutes; [distributed by Jessie Heath, 1142 Thornton Road, Houston, TX 77018-3233; http://home.earthlink.net/~jheath1821].

2000 Maligiaq Makes Waves on His U.S. Visit. *Sea Kayaker* 17(2): 55–61.

Petersen, H.C.

1986 *Skinboats of Greenland*. Trans. K. M. Gerould. Ships and Boats of the North, vol. 1. Roskilde: National Museum of Denmark, the Museum of Greenland, and the Viking Ship Museum.

Silis, Ivars (director)

1995 *Amphibious Man*. Video directed by Ivars Silis; produced by KNR/Greenland, TV2/Denmark and Ivars Silis; 28 minutes; [distributed by Jessie Heath, 1142 Thornton Road, Houston, TX 77018-3233; http://home.earthlink.net/~jheath1821].

Stamer, Greg

1998 Greenland Technique from the Source: Lessons Learned from Maligiaq Padilla. *ANorAK* (November/December) 16(6):6–9.

2000 The Petrussen Maneuver: A New Twist on an Old Technique. *Sea Kayaker* 17(3):15–17.

2001 2000 Greenland National Kayaking Championship. *Sea Kayaker* 17(6):34–45.

2003 Allunaariaqattaarneq–Inuit Rope Gymnastics. *Sea Kayaker* 19 (6):47–53.

Wilkes, R. Jeffrey

1999 Water Wings. Retrieved May 11, 2003 from http://www.phys.washington.edu/~wilkes/post/temp/phys208/notes/lect19.html.

3

Kayaks in European Museums: A Recent Research Expedition

HARVEY GOLDEN

The research presented here comes from a trip conducted in 1998 to museums in England, Scotland, and the Netherlands as well as from a trip to Nuuk, Greenland, in 2000. Eleven of the 38 kayaks surveyed during these trips have been selected for this chapter; they are described and presented in scale lines drawings.

The kayaks represent a range of some 300 to 400 years of development. Geographically, they range from Polar Greenland to West and East Greenland. They exhibit a large range of forms, sizes, and proportions, thereby emphasizing a diversity in kayaks so often lumped together as "Greenland kayaks."

Two published sources were fundamental in this project. One is William Clark Souter's *The Story of Our Kayak and Some Others*, which provided a list of 33 kayaks throughout the United Kingdom (1934:13). I surveyed nine of these kayaks and observed others on the 1998 trip. (John Brand has also surveyed several of the kayaks on this list; see excerpt, this volume.) Tracking down kayaks from the 1934 list was tricky. Six or more kayaks from the list have been lost or destroyed. War brought the end to a few kayaks in Britain: naval commandos apparently sea-tested actual museum kayaks in their search for designs suitable for military use, according to Robert Keith Headland (personal communication, 1998), formerly archivist and curator at the Scott Polar Research Institute, Cambridge.

The second source is a more recent publication and covers kayaks in the Netherlands. Gert Nooter's "Old Kayaks in the Netherlands" (1971), lists 18 kayaks, but focuses on 10 kayaks thought to date from circa 1600–1800 that all hail from West Greenland. I surveyed five of the old kayaks and six of the others on the 1998 trip. I was able to observe a couple of the others on Nooter's list, but was unable to take lines due to their inaccessibility. Nooter himself collected several of the East Greenland kayaks on the list, two of which appear here.

West Greenland Kayak, c. 1650–1750 ("Hoorn"). This kayak (Figure 3.1) is described at length by Gert Nooter in *Old Kayaks in the Netherlands*, and he is as thorough as possible in regards to early data. Unfortunately, the earliest source of information is an entry in a Westfries Museum catalog from 1890 (Nooter 1971:24).

The tradition associated with this kayak is that it was found adrift in the North Sea—with a body inside (Nooter 1971:24). It's conceivable that it may be one of those used by Greenlanders fleeing from Denmark, or perhaps from a ship that brought them from Greenland. Isaac de la Peyrère describes several successful as well as failed escapes from Denmark in the early 1600s (Peyrère 1855:221–227). "Success" is relative, of course, for surely none made it even close to Greenland, and the kayakers undoubtedly met death on the seas or shores of northern Europe.

The average width of the ten "old kayaks" in Nooter's book is 16⅕ inches (41.9 cm) (Nooter 1971:74). The Hoorn kayak is one of the ten from the list, and is almost 4¾ inches (12 cm) wider than this average. Such an anomaly suggests certain possibilities about the intended use and user of this kayak. Perhaps the user was of poor balance or a beginner, or perhaps the kayak was intended primarily for fishing or another function requiring more stability. It is extremely unlikely that the Hoorn kayak was made for a larger person: the cockpit opening is very small (15¾ inches or 40 cm) long by 14 inches (35.6 cm) wide and the kayak is quite shallow (6½ inches [16.5 cm] depth-to-sheer).

Figure 3.1
WEST GREENLAND KAYAK, c. 1650–1750
Westfries Museum, Hoorn, the Netherlands
no. 0232.1
Surveyed by Harvey Golden, 1998

Length	18'2"	553.7 cm
Beam	21³⁄₁₆"	53.8 cm
Depth to sheer	6³⁄₈"	16.2 cm
Depth overall	8¹⁄₁₆"	20.4 cm

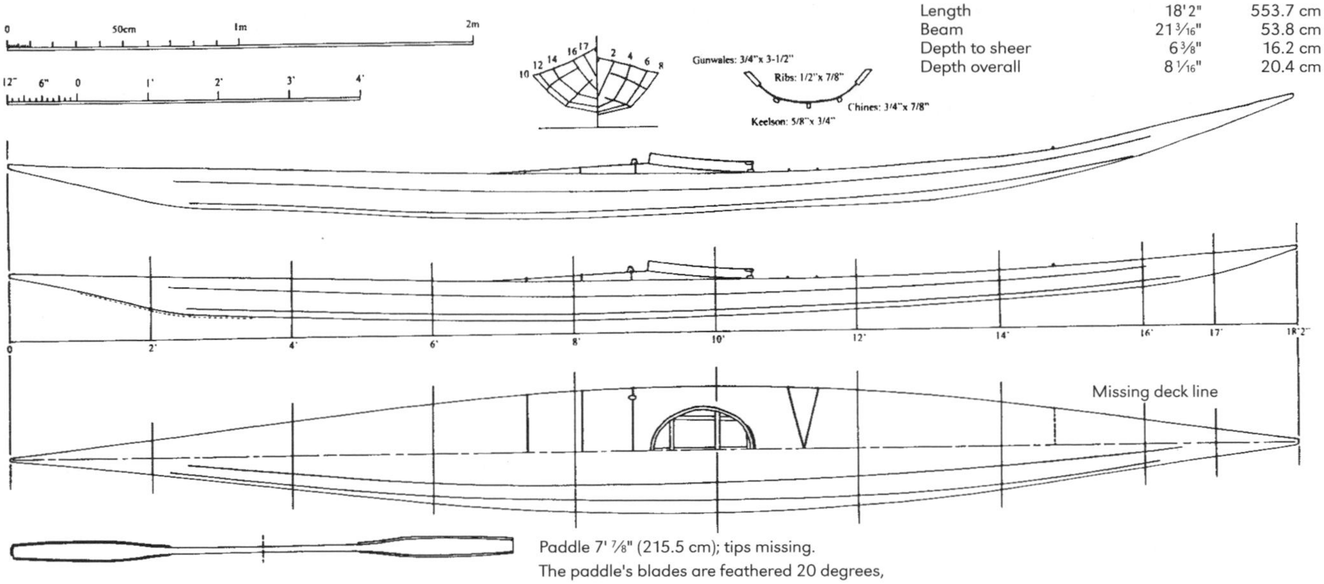

Gunwales: 3/4"x 3-1/2"
Ribs: 1/2"x 7/8"
Chines: 3/4"x 7/8"
Keelson: 5/8"x 3/4"

Missing deck line

Paddle 7' ⅞" (215.5 cm); tips missing.
The paddle's blades are feathered 20 degrees,
but are drawn flat.

According to Kaj Birket-Smith, "There may be some, if only rather a slight difference in the proportions of the kayak, according as [*sic*] the chief occupation of the owner is seal hunting or fishing" (1924:265). Birket-Smith goes on, citing H. P. Steensby (1912:164): "Fisherman are as a rule apt to make their kayaks broader and more flat-bottomed. Steensby connects this with the fact that the fishermen as a rule are less capable kayakers than the seal hunters. This is incontestable, but it must also be borne in mind that the fisherman has greater opportunity of lying still, and this is most easily done with a comparatively broad kayak" (1924:265–266).

The builder of the Hoorn kayaks has paid great attention to ensuring that an abnormally wide kayak would paddle very much like its narrower counterparts in certain rougher conditions. A proportional yet wider kayak will inherently result in a larger volume, but the builder of the Hoorn kayak countered this result by decreasing the depth and making the sides greatly flared. H. C. Petersen describes the importance of properly balanced volume: a kayak "must have a certain volume so that it will not sink too quickly in the water. But it must not ride too high in the water, either. With too

little air it would be difficult to rise above storm waves. Too much air makes it ride high in the water and increases the wind pressure and drift of the kayak" (1986:42).

The upper parts of the Hoorn kayak are in good condition, but some sections of the bottom have collapsed over the years. There are a number of arched crosspieces as well as an empty mortise above a lower crosspiece that likely doubled as a footrest. Just forward of this is a long clamping tie drawing the lower gunwale edges in. Broken ribs and debris are just forward of this clamping tie. The deck stringers have slipped out of place. Another heavy clamping tie helps hold the flare in the gunwales. Cluttered debris and loose deck beams obscure the stern of the kayak.

Bone or ivory edging is in evidence along the keelson at the bow. The only other evidence of bottom protection is two sets of holes along the left chine, the longitudinal stringer adjacent the keelson: the first two holes are at 6 feet, 4 inches (193.0 cm) and 6 feet, 4⅞ inches (195.3 cm) from the bow; the third and fourth are at 10 feet, 6 inches (320.0 cm) and 10 feet, 8 inches (325.1 cm). The pieces were situated over transverse seams (areas prone to high wear), but strangely, they appear only on the left chine. A deck line is also missing at 14 feet, 9 inches (449.6 cm) from the bow. There is no evidence of end knobs, though part of the stern seems to be missing.

The paddle of this kayak is very well made, though its tips are missing. It is slightly feathered, the blades angled 20 degrees from each other. John Brand describes a feathered Greenland paddle (1987:47) associated with the South Shields Museum kayak. This was the first feathered Greenland paddle thoroughly studied, and at the time it was not known for certain whether others existed. It has been suggested that such paddles are simply warped, but they may also have been carved from twisted pieces of wood, using such grain to advantage. If it had twisted, the paddle showed no further warps, and was straight and true along its longitudinal axis. The ivory edging on the paddle is ⅛ inch by ⅛ inch (3 mm).

Save for its being feathered, the Hoorn kayak's paddle is nearly identical in shape to the example Hugh Collings surveyed at Skokloster Castle. The sheer line of the Hoorn kayak is very similar to that of the Skokloster, and they likely hail from close vicinity. Nooter suggests that the Hoorn kayak was from the Sukkertoppen/Holsteinborg area of West Greenland (1971:24). This conclusion is to be held as suspect, since it was based on a typology (Birket-Smith 1924:270–271) developed primarily in the early 20th century; such a conclusion assumes that kayak form had been more or less static for some 300 years.

A harpoon line stand is also associated with the kayak. This stand suggests that, while the kayak may have been designed primarily for fishing, it was also used for sea-mammal hunting. The stand is symmetrically constructed, unlike later types that have one leg protruding diagonally from the center of the rack. The rack is missing the transverse crosspieces that support the line, but is otherwise in good condition. If indeed the rack was found with the kayak, it would be likely that the fleeing kayaker had intended to hunt en route.

West Greenland Kayak, c. 1600–1700 ("Hindeloopen"). This kayak (Figure 3.2) has extensive hull collapse, mainly concentrated behind the cockpit. Planking had been inserted in the cockpit area to maintain rigidity—likely an early attempt at conservation. Tin tips cover the bow and stern of this kayak; their purpose is mysterious, but they were likely attached by non-Greenlanders. It is on display in a case at De Hidde Nijland Stichting under extremely heavy plate glass. When Nooter visited, the Hindeloopen kayak was suspended from the ceiling (1971:18–21).

Nooter cites two sources, both from 1919, that offer conflicting information on the history of the kayak: one reports that it was collected by Hindeloopen whalers at "Iseland," and the other states that it had washed ashore at Hindeloopen "at the beginning of this century" (1971:18, 21). The latter suggestion could be discounted, as the kayak certainly predates 1900.

The Hindeloopen kayak has a wavy waterline painted on it—evidently patterned after 17th-century Dutch ships. The wave pattern is white and descends to just below the chines. The upper part of the kayak is painted in a very dark green, nearing black. The coaming was also painted this color as was the paddle associated with the kayak. Two other kayaks in the Netherlands have this same paint scheme: the child's kayak in Hollum and the kayak in De Rijp (Nooter 1971:15, 38).

The forward-most deck line is missing; it may or may not have had ivory or bone fittings on it. A single line just aft of the cockpit is also missing. Bow ice protection is evidenced by ⅛-inch holes for attaching cutwater edging every inch or so to about 3 feet (91.4 cm) aft of the bow. No stern edge protection is evidenced.

West Greenland Kayak, c. 1600–1700 ("Brielle"). Nooter also features this kayak (Figure 3.3) in his 1971 volume; he refers to it as the "Brielle," after the Dutch town it had been in before being turned over to the Rijksmuseum voor Volkenkunde in Leiden in 1883 (1971:61–62). As with the Hindeloopen kayak, little is known about the history of the Brielle kayak. Nooter suggests that it originates from around Disko Bay (1971:61) based on Birket-Smith's kayak typology (1924:271). Again, this suggestion is highly suspect due to its having been developed based on kayak forms of the early 20th century.

The Brielle kayak is in a very damaged state: viewed from above, it winds like a snake. Extensive hull collapse exists, primarily in the stern. The interior is cluttered with broken deck beams, chines, and ribs; the keelson is missing at the cockpit area. The skin is in fair shape and clean. It is a dark rusty-brown color. The drawing (Figure 3.3) presents a reconstructed kayak—drawn with a straight centerline, as it was certainly built. Likewise, the elevation (Figure 3.3) depicts a conjecturally restored profile, the hull collapse being faired out. The sheer line is depicted as surveyed, in very good condition.

Note that some of the photographs of kayaks in Nooter's book are erroneously labeled: figure 21 (Nooter's number 15) is the Brielle, *not* the Hague, and Nooter's figure 28 is the Hague (Nooter's number 16), *not* the Brielle. Nooter mentions two "Hague" kayaks: one, number 16 (originally Rijksmuseum voor Volkenkunde catalog no. 351-78), has since been transferred to the Greenland National Museum, Nuuk.

Figure 3.2

WEST GREENLAND KAYAK, c. 1600–1700
Provenance unknown
De Hidde Nijland Stichting, Hindeloopen,
    the Netherlands, no. 2
Surveyed by Harvey Golden, 1998

Length	18'8"	568.9 cm
Beam	15 5/8"	39.7 cm
Depth to sheer	6 7/8"	17.4 cm
Depth overall	8"	20.3 cm

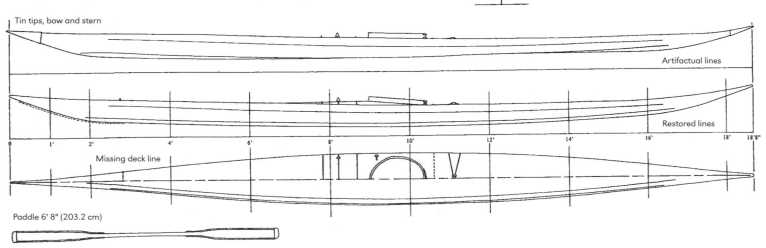

Tin tips, bow and stern

Artifactual lines

Restored lines

Missing deck line

Paddle 6' 8" (203.2 cm)

Figure 3.3

WEST GREENLAND KAYAK, c. 1600–1700
Rijksmuseum voor Volkenkunde, Leiden,
    the Netherlands, no. 349-1 (formerly located in the
    town of Brielle, the Netherlands)
Surveyed by Harvey Golden, 1998

Length	17'11 1/8"	546.4 cm
Beam	15 1/4"	38.7 cm
Depth to sheer	6 1/4"	15.8 cm
Depth overall	7 1/8"	18.1 cm

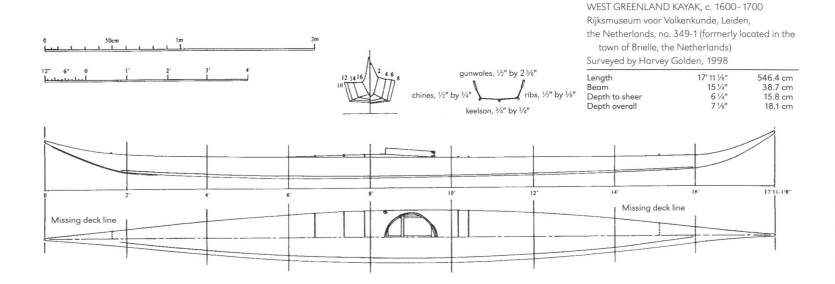

gunwales, 1/2" by 2 5/8"
chines, 1/2" by 3/4"   ribs, 1/2" by 3/8"
keelson, 3/4" by 3/4"

Missing deck line

Missing deck line

Missing deck line

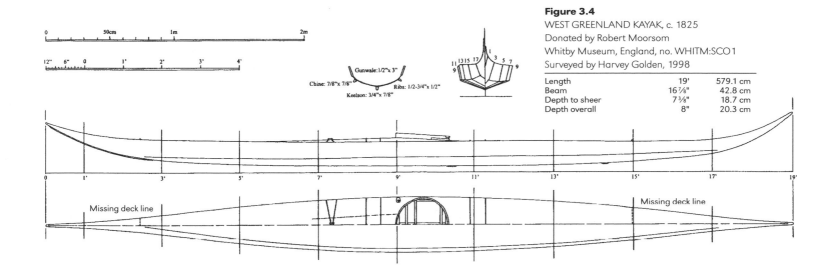

Figure 3.4
WEST GREENLAND KAYAK, c. 1825
Donated by Robert Moorsom
Whitby Museum, England, no. WHITM:SCO1
Surveyed by Harvey Golden, 1998

Length	19'	579.1 cm
Beam	16⅞"	42.8 cm
Depth to sheer	7⅜"	18.7 cm
Depth overall	8"	20.3 cm

Gunwale:1/2"x 3"
Chine: 7/8"x 7/8"   Ribs: 1/2-3/4"x 1/2"
Keelson: 3/4"x 7/8"

Missing deck line

Missing deck line

West Greenland Kayak, c. 1825. A Richard Moorsom donated this kayak (Figure 3.4) to the Whitby Literary and Philosophical Society in North Yorkshire in 1825. Whether or not Moorsom was the collector is not known. The kayak could be much older than 1825, as Whitby had been sending whalers north since the mid-18th century.

The Whitby kayak has especially long and thin ends, reminiscent of Caribou Inuit kayaks of central Canada. Also noteworthy and further reminiscent of central Canadian kayaks is the reverse sheer, peaking just aft of the cockpit. This is very subtle yet again suggests a not-too-distant lineage to central Canadian kayaks. Downward warping of the gunwales has been considered as the cause, but upon examination of the keel and chines, this was dismissed, as they are not collapsed or deformed, and in fact continue along a fair and consistent curve.

While the Whitby kayak resembles the Brielle (Figure 3.3), the cross sections are fundamentally different, the former having significant dead rise while the latter is very flat-bottomed.

The cross section of the Whitby kayak is remarkable in that it has a very steep angle of dead rise and the chines are placed quite high. Such a deep V bottom is not common among preserved West Greenland kayaks from this period and earlier, though it slowly became the standard later in the 19th century. (One other exception is the 17th-century Wrangel Armoury kayak described by Hugh Collings, this volume.) Petersen writes of the advantage of V-bottom kayaks: "it is

of course more difficult to maintain one's balance but the shape does give the kayaker more control over his craft" (1986:46).

With such a prominent keel, this kayak would presumably track well, though its considerable rocker ought to provide for good turning, as is the case with the Wrangel Armoury kayak replica built and paddled by the author.

The fittings on this kayak are simple in having only one toggle aside from the harpoon holder. Two deck lines are missing: the farthest forward (which may have held ivory knobs), and the farthest aft. Bow ice protection is also depicted, as are two small pieces aft that cover the transverse seams of the skins. Stern ice protection was not noted, but there was evidence of bow and stern knobs. There are no aft deck stringers on this kayak, which is somewhat unusual for West Greenland kayaks.

West Greenland Kayak, c. 1789. Museum catalog information shows that a Mr. Watson, commander of the Greenlandman *Findlay*, collected this kayak in the Davis Strait region in 1789. The Glasgow Whale Fishing Company presented it to the University of Glasgow in 1790. The Hunterian Museum has three paddles (and two other kayaks), but it is not known which paddle belongs to this kayak, though one is arbitrarily presented with the lines drawing (Figure 3.5).

With its high ends and great sheer, this kayak is the type described by H. C. Petersen as an *avasisaartoq*, the curved type (Figure 3.6). Petersen writes, "There are many variations of the *avasisaartoq* kayak.

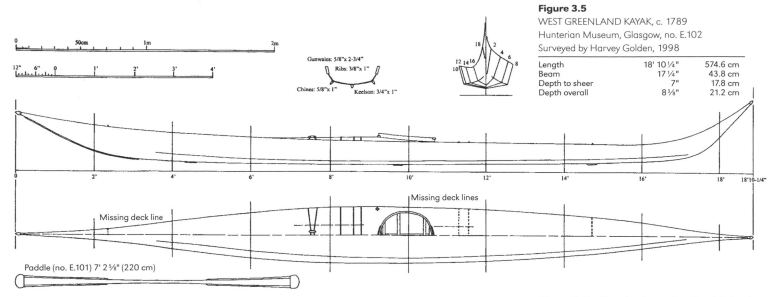

Figure 3.5
WEST GREENLAND KAYAK, c. 1789
Hunterian Museum, Glasgow, no. E.102
Surveyed by Harvey Golden, 1998

Length	18' 10 ¼"	574.6 cm
Beam	17 ¼"	43.8 cm
Depth to sheer	7"	17.8 cm
Depth overall	8 ⅜"	21.2 cm

Gunwales: 5/8"x 2-3/4"
Ribs: 3/8"x 1"
Chines: 5/8"x 1"
Keelson: 3/4"x 1"

Missing deck lines

Missing deck line

Paddle (no. E.101) 7' 2 ⅝" (220 cm)

Figure 3.6—Greenlandic *avasisaartoq*-style kayak, c. 1870. Photographer W. Bradford. Courtesy National Archives of Canada, no. 81978.

It seems to have been widespread until the end of the 19th century. After the introduction of the rifle it became necessary to straighten out the sharply rising ends and after that the type quickly disappeared from the areas where it had formerly been in use" (1986:49).

This kayak is in excellent condition for its age. The skin is a clean, creamy yellow color and the coaming is of a bright reddish-colored wood. Both end knobs are present, as is the keel strip at the bow.

Several pieces of ivory dot the keelson where there are transverse seams. The foremost deck line is missing, as are all the deck lines aft of the cockpit. The pair behind the cockpit may have had a toggle that joined them, much like the example forward of the cockpit.

The deck beams are not lashed to the gunwales, and only one deck beam in the front is pegged. Likewise, only one deck beam in the aft half of the kayak is pegged. The cross piece at the lower edge of the gunwales toward the bow holds the gunwales together and helps set the gunwale flare. The deck stringers are shimmed to the correct height with small wooden squares.

West Greenland Kayak, c. 1889. This is a child's kayak (Figure 3.7) from before 1889, provenance unknown, though it is likely from the mid-to-upper west coast. For a kayak whose young owner would quickly outgrow it, it is extremely well made, no detail compromised. The kayak's lines are graceful and the sheer has substantial curve over its 8-foot-5-inch (256.5 cm) length. The bow is moderately high, and the stern, though short and steeply raked, rises up nicely at the tip.

There are three ivory pieces on the foremost deck line, and a harpoon rest is present adjacent the cockpit. Keel edging and end knobs, both fore and aft, are present. A spray skirt is attached to the kayak's coaming and is depicted in the scale drawing.

The museum data card describes the kayak as having been made for a six- or seven-year-old. It is attractive and must have been the source of enormous pride for both teacher and the young kayaker.

West Greenland Kayak, mid to late 19th century. Captain Thomas Robertson of the whaler *Scotia* collected this kayak in 1908 (Figure 3.8). Museum records do not record any data aside from its origin— the "Great Northern River." (Research into what Scottish whalers meant by this is warranted; it may refer to Davis Strait or perhaps to a specific vicinity on the coast of Greenland.)

While the 1968-81-1 is clearly an *avasisaartoq*, as is the Hunterian Museum's E.102, the ends of the former are far more restrained in form. Based on Kaj Birket-Smith's kayak typology (1924:271), this kayak possibly originated from Umanaq Fjord or Disko Bay vicinities.

Ice protection along the keelson exists at the bow as well as the stern—in both cases extending all the way up to the end knobs. The harpoon holder is missing but is evidenced on the front right quarter of the cockpit. The forward-most deck line is also missing. The farthest forward line seems to be missing in most museum kayaks—likely because they were used, improperly, to carry the kayak.

The paddle is positively associated with the kayak. The end pieces are fairly long; the edging is ¼-inch (6 mm) thick and ⅜-inch (1 cm) tall at the ends, tapering to ⅛ inch (3 mm) toward the middle section of the paddle.

West Greenland Kayak, mid-20th century. The history of this kayak (Figure 3.9a, b) was uncertain, though Peqatigiiffik Qajaq Nuuk (PQN) secretary Hans Kleist-Thomassen attributed it to the 1950s (personal communication). It is likely a type from the vicinity of Nuuk. The PQN frame hangs inside the kayak clubhouse on the colonial harbor.

The condition of this kayak frame is very good, although it is missing a few ribs and deck beams. It has a very attractive sheer line and a deep bow. The stem and stern are considerably long, and each has a subtle yet elegant concavity in its profile.

A variety of deck beams/gunwale joints are exhibited with this kayak. Most of the beams are lashed to the gunwales, but two deck beams exhibit the oblique pegging method. One of these beams is missing, but the other is intact; the pegs do not go all the way through the gunwales. A couple of deck beams are missing, as are four ribs. The fore- and aftmost deck beam mortises and the foremost rib mortise were likely never used.

The ribs are pegged through their tenons, though several have been glued in place with polyurethane glue—likely a recent repair. The keelson of the PQN kayak is made from two pieces, the scarf being of a simple stepped form, nailed in place. Where the wider keelson ends meet the gunwales, a "key" block is used to restrict shifting.

At the forward keelson-to-gunwale joint, a shim is used to build up extra height. Whether or not the keelson was recycled from a different kayak is unknown. All of the lashings are either heavy nylon fishing line or natural twine.

Another interesting thing to note about the PQN is the *ittivik* (rear cockpit deck beam) and its placement. It is mortised and tenoned in place, but the mortises do not go all the way through. The *ittivik* is also about ⅜ inch (1 cm) lower than the deck beam behind it. It is likely that it was added as a modification; the beam presently behind the *ittivik* was likely the *ittivik* at one time. The obvious result of this is a much shorter cockpit that is only 13¼ inches (33.7 cm) long.

A second scale drawing of this kayak (Figure 3.9b) is also presented: an "outside lines" rendering, thereby highlighting this kayak's form.

East Greenland Kayak, c. 1933. This kayak (Figure 3.10) was acquired by N. Tinbergen at Ammassalik, East Greenland (Map 2), in 1932 or 1933, and is one of two kayaks he brought back with him to the Netherlands. The other was made for his use and is at the Rijksmuseum voor Volkenkunde, Leiden. These kayaks are from the same period and place as those used by the British Arctic Air Route Expedition.

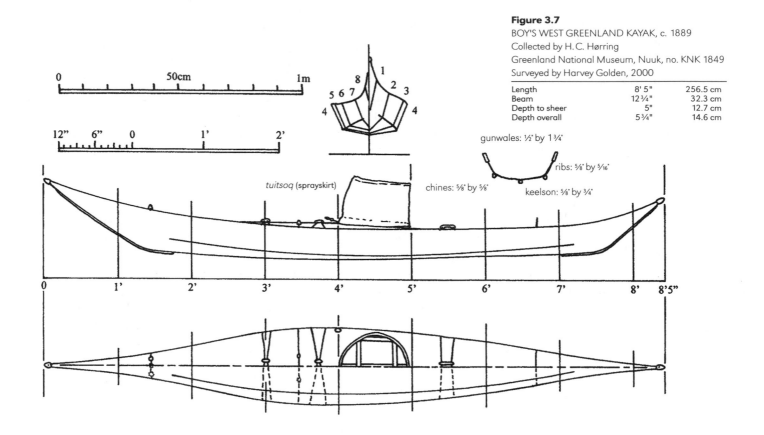

Figure 3.7
BOY'S WEST GREENLAND KAYAK, c. 1889
Collected by H.C. Hørring
Greenland National Museum, Nuuk, no. KNK 1849
Surveyed by Harvey Golden, 2000

Length	8' 5"	256.5 cm
Beam	12¾"	32.3 cm
Depth to sheer	5"	12.7 cm
Depth overall	5¾"	14.6 cm

gunwales: ½" by 1¾"

ribs: ⅝" by 5⁄16"

chines: ⅝" by ⅝"

keelson: ⅝" by ¾"

tuitsoq (sprayskirt)

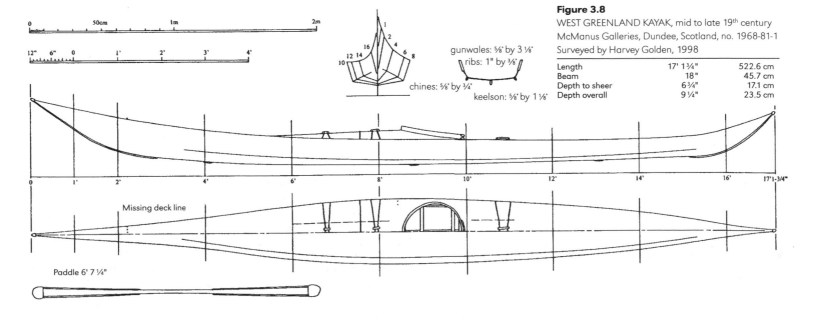

Figure 3.8
WEST GREENLAND KAYAK, mid to late 19th century
McManus Galleries, Dundee, Scotland, no. 1968-81-1
Surveyed by Harvey Golden, 1998

Length	17' 1¾"	522.6 cm
Beam	18"	45.7 cm
Depth to sheer	6¾"	17.1 cm
Depth overall	9¼"	23.5 cm

gunwales: ⅝" by 3⅛"

ribs: 1" by ⅜"

chines: ⅝" by ¾"

keelson: ⅝" by 1⅛"

Missing deck line

Paddle 6' 7¼"

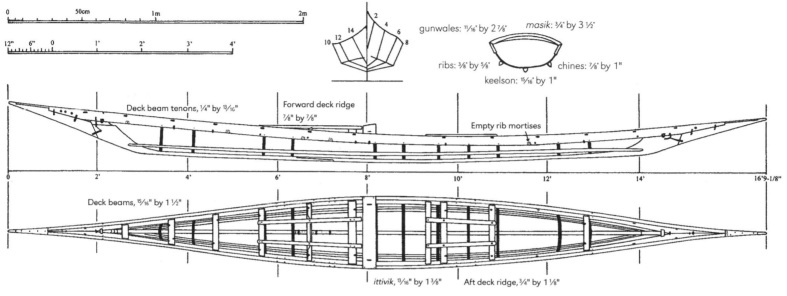

Figure 3.9a

WEST GREENLAND KAYAK, mid-20th century

Nuuk Kayak Club Collection

Surveyed by Harvey Golden, 2000

Length	16' 9 ⅛"	510.8 cm
Beam	19 ⁷⁄₁₆"	49.3 cm
Depth to sheer	7 ⅝"	19.3 cm
Depth overall	10 ³⁄₁₆"	25.8 cm

0   50cm   1m   2m

12" 6" 0 1' 2' 3' 4'

gunwales: ¹¹⁄₁₆" by 2 ⅞"   *masik*: ¾" by 3 ½"

ribs: ⅜" by ⅝"   chines: ⅞" by 1"

keelson: ¹⁵⁄₁₆" by 1"

Deck beam tenons, ¼" by ¹³⁄₁₆"   Forward deck ridge ⅞" by ⅞"   Empty rib mortises

0   2'   4'   6'   8'   10'   12'   14'   16'9-1/8"

Deck beams, ¹⁵⁄₁₆" by 1 ½"

ittivik, ¹³⁄₁₆" by 1 ⅜"   Aft deck ridge, ¾" by 1 ⅛"

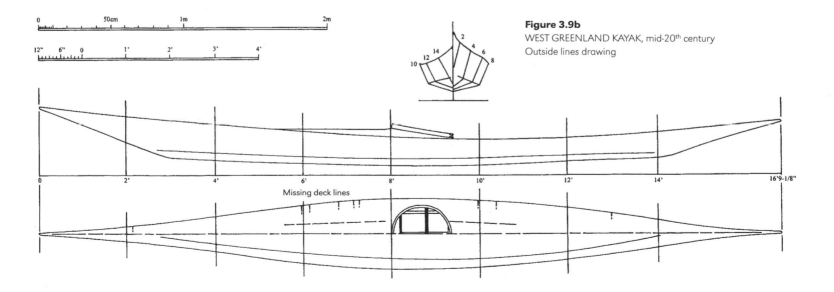

Figure 3.9b

WEST GREENLAND KAYAK, mid-20th century

Outside lines drawing

0   50cm   1m   2m

12" 6" 0 1' 2' 3' 4'

0   2'   4'   6'   8'   10'   12'   14'   16'9-1/8"

Missing deck lines

The 48057 is not only representative of the height of East Greenland kayak design but is also an especially fine example of workmanship and skill. Not a single element of this kayak is poorly made; only the hastily repaired paddle tip stands out as less than perfect. Even the equipment associated with the 48057 is well made: the ivory-plated harpoon line stand (*asaloq*), the intricately scrimshawed harpoon blade, and the carved and polished fittings. Even the mundane paddle holder—a flat batten of wood—has carved ivory seals pegged to it.

The cross section of this kayak contrasts dramatically with the West Greenland kayaks. The narrow, flat bottom, steeply angled sides, and shallow hull combine to create an almost dish-shaped form offering very good stability and minimal freeboard.

The 48057 is very dark, almost black in color, and the oil in the skin is still tacky and odiferous. All of the deck fittings are made of ivory. A wooden drain plug is present on the aft starboard side of this kayak. This kayak is well protected from ice: long strips of ivory or bone protect the stems and intermittent pieces protect raised seams and other high-wear areas. The chines are also intermittently protected with ivory or bone strips. The stern knob is missing, but evidenced.

The gunwales have considerable flare, and all the deck beams are pegged fast to the gunwales. There are three wide forward deck stringers. This kayak has no aft deck stringers, nor does it have heel guides (as seen in Figure 3.11). The *masik* (arched cockpit deck beam) has a carved handle, which facilitated entry into snug East Greenland kayaks.

Inside, three wide deck stringers provide a sturdy platform for the hunting implements. The dark cords that pass through these stringers are the continuous deck lines. The deck beams are all pegged and mortised into place; one deck beam is missing. The ribs are nailed to the chines and keelson.

East Greenland Kayak, late 1960s. Henrik Singhertek of Tiniteqilaaq built this kayak (Museon no. 59876) in Ammassalik district (Figure 3.11). It receives brief mention in Nooter (1971:1–3) when he presents a list of old and newer kayaks. Nooter collected the kayak himself in 1970 during fieldwork in East Greenland. He has described East Greenland kayaks in detail elsewhere (1991).

This kayak is shorter than the kayak in Figure 3.10, and was likely used for hunting in especially icy waters; the short length facilitated dogsledding to leads. The 59876 has fewer and much simpler deck fittings, though only one is actually missing, as evidenced by two holes along the sheer. (The placement of the missing lines suggests that it was the Y-shaped adjustable line used to hold the harpoon line stand in place. A similar such line can be seen in Figure 3.10.)

The 59876 features two drain plugs instead of the usual single plug. The plug not shown in the drawing is on the other side (starboard) and is an inch farther forward than its visible counterpart.

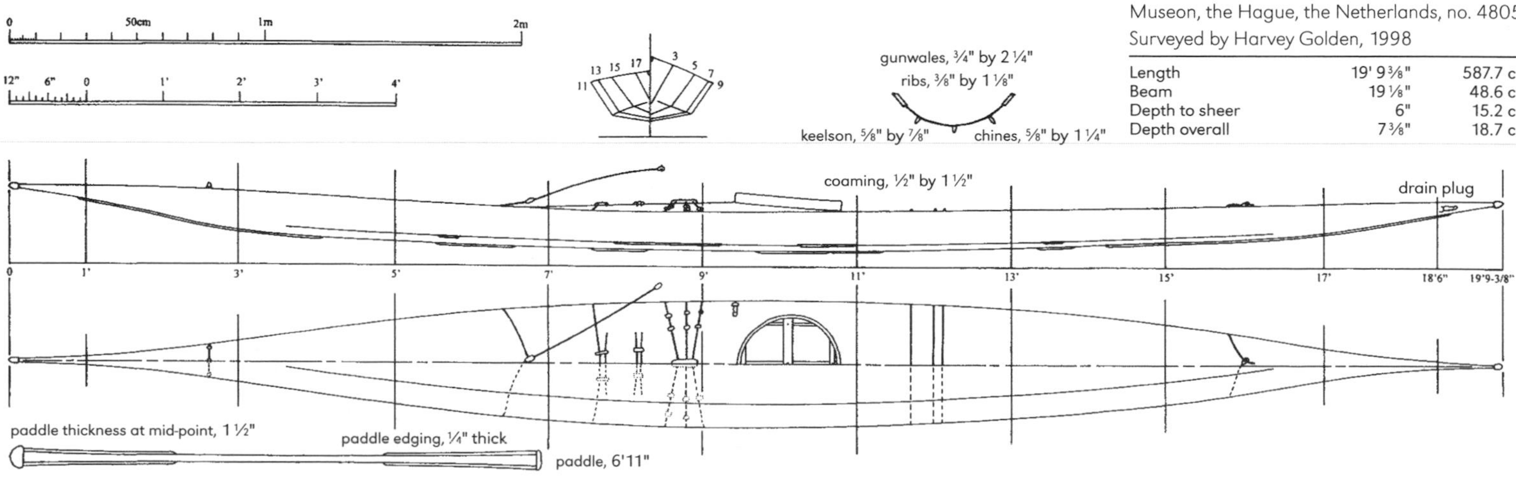

Figure 3.10
EAST GREENLAND KAYAK
Collected in Ammassalik, 1932–1933
Museon, the Hague, the Netherlands, no. 48057
Surveyed by Harvey Golden, 1998

Length	19' 9 3/8"	587.7 cm
Beam	19 1/8"	48.6 cm
Depth to sheer	6"	15.2 cm
Depth overall	7 3/8"	18.7 cm

gunwales, 3/4" by 2 1/4"
ribs, 3/8" by 1 1/8"
keelson, 5/8" by 7/8"   chines, 5/8" by 1 1/4"
coaming, 1/2" by 1 1/2"
drain plug
paddle thickness at mid-point, 1 1/2"
paddle edging, 1/4" thick
paddle, 6'11"

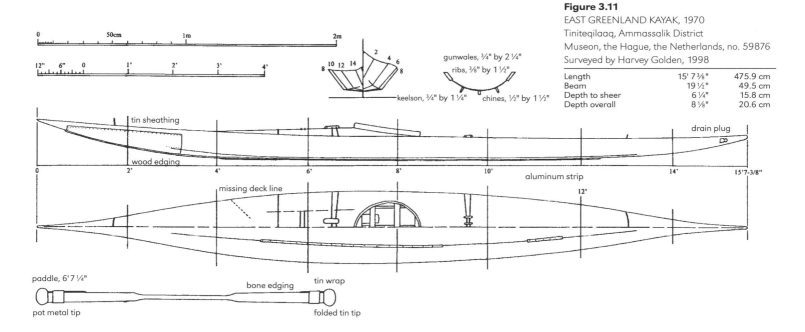

Figure 3.11
EAST GREENLAND KAYAK, 1970
Tiniteqilaaq, Ammassalik District
Museon, the Hague, the Netherlands, no. 59876
Surveyed by Harvey Golden, 1998

Length	15' 7⅜"	475.9 cm
Beam	19½"	49.5 cm
Depth to sheer	6¼"	15.8 cm
Depth overall	8⅛"	20.6 cm

gunwales, ¾" by 2¼"
ribs, ⅜" by 1½"
keelson, ¾" by 1¼"   chines, ½" by 1½"

tin sheathing
wood edging
drain plug
aluminum strip
missing deck line

paddle, 6'7¼"
bone edging
tin wrap
pot metal tip
folded tin tip

Most unusual yet appropriate is the tin armor this boat wears: the entire forward section below the sheer is sheathed in metal to protect it from ice damage. It is tacked on through the sealskin covering along the gunwale, chine, and keelson. A photograph in Nooter's 1991 work shows how this would be of great advantage (1991:336, fig. 20).

While all the chine and keel protection on the 48057 is ivory or bone, the 59876 has wood and aluminum strips. Like the 48057, this kayak has three wide forward deck stringers and no aft deck stringers.

The deck fitting just forward of the cockpit has three holes drilled into it, but only two deck lines go through this piece. There is no evidence of a deck line missing adjacent to the existing ones, so we can assume that this fitting has been recycled from an older kayak. The 48057 has three lines through its corresponding fitting.

The cockpit layout of the 59876 contains a fiberboard seat; the seat slats are nailed on from beneath. The ribs are double-nailed to the chines and keelson, and on the left, a hide shim has been placed to ensure fair lines. Ivory nails along the inner coaming support the skin cover, which have come loose on the front edge. The *masik* has a carved handle in it; this facilitated not only lifting the kayak but also getting into it.

The paddle with the 59876 also utilizes modern materials: the ends are of metal—one is a chunk of pot metal, the other a folded and riveted aluminum sheet. Only two small pieces of bone protect the paddle's edges, and they do so on just one edge. These pieces are toward the ends, and have been further reinforced with a tin wrapping that goes all the way around the blades. Much as the kayak is armored, so is its paddle.

Polar Greenland Kayak. Thomas Thomsen of the Danish National Museum collected this kayak in 1909 (DNM catalog no. L.4348). In 1990, it was transferred from the Danish museum to the collection of the Greenland National Museum, Nuuk. The history of the kayak among the Polar Greenland Eskimo is an interesting one. Their ability to build kayaks had been entirely lost, though the concept and the word "qayaq" still remained. This knowledge and skill was reintroduced in the early 1860s after the arrival of immigrants from North Baffin Island (Steensby 1910:261), and has given the Polar Greenlanders a kayak with distinctly Canadian elements.

The KNK 1007 (Figure 3.12) has a swede-form hull (i.e., carrying its beam well aft of the cockpit, with a narrower bow), single deck ridge, clipper bow, and very shallow stern. The KNK 1007 has pieced ribs (i.e., not bent), which gives it a flat-bottomed hull. The coaming is also pieced together and is more or less triangular in form. No baleen seems to have been used in the construction or lashing of the 1007; the Polar Greenlanders did not hunt the larger baleen whales (Gilberg 1984:582; Rasmussen 1921:12–13).

The deck lines of the 1007 are very simple. The forward-most strap is slack and fastened through the skin and gunwales. The straps ahead of the cockpit are not paired fore and aft, but instead are paired left and right—tied together in the middle with a separate cord. Tied to the paired cords is a large chunk of whalebone, crudely fashioned into a wedge shape; this piece served as a harpoon rest. Behind the cockpit is a single line, with a bit of slack in it. Beyond this, there is evidence of a line having been about two feet ahead of the stern.

The paddle associated with the 1007 is smooth and very well made, although it is not entirely symmetrical. One of the paddle's blades has been made wider by a piece of wood tied on with lashings. The blades are still of unequal sizes, the one with the lashed-on piece being larger. A leather cord is wrapped once around one blade, set in tiny notches at the edges. This cord functioned as a drip ring, stopping cold water from running down the blade onto the paddler's hands. The other wrap is missing, but the notches are evident. The

knot on the existing wrap is situated right at the blade's edge and its tails dangle a half-inch or so.

An unusual feature of the Polar Greenland kayak is the large semicircular piece of whalebone attached to the front piece of the coaming. Apparently this was used as a paddle rest—not just when the paddler was resting but while under way as well.

Polar Greenland kayaks have inspired an unusual range of commentary by European and American scholars. Naval architect and historian Howard Chapelle writes quite favorably of these kayaks: "[They] are highly developed craft—stable, fast, and seaworthy—and the construction is light yet strong enough to withstand the severe abuse sometimes given them" (Adney and Chapelle 1964:206). Ethnologist H. P. Steensby wrote in 1910, after his visit to Polar Greenland, that the kayaks there were "clumsy, heavy, and open" (Steensby 1910:290). Steensby's description is more accurate—especially as he is comparing them to more southerly Greenland kayaks, though Chapelle's observation is not untrue: these kayaks were used quite ably, and certainly met the needs of those relying on them.

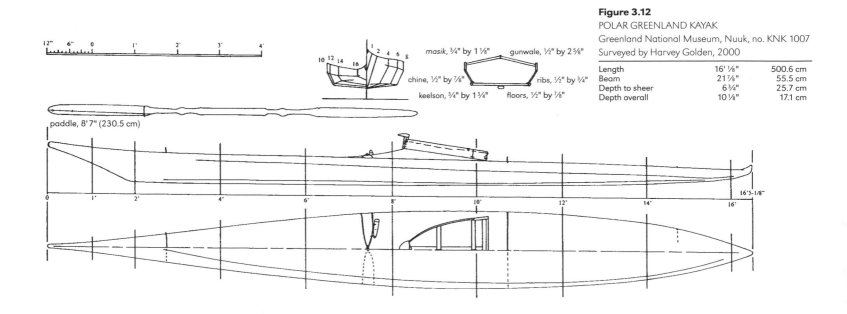

Figure 3.12
POLAR GREENLAND KAYAK
Greenland National Museum, Nuuk, no. KNK 1007
Surveyed by Harvey Golden, 2000

Length	16' 1/8"	500.6 cm
Beam	21 7/8"	55.5 cm
Depth to sheer	6 3/4"	25.7 cm
Depth overall	10 1/8"	17.1 cm

masik, 3/4" by 1 1/8"   gunwale, 1/2" by 2 5/8"
chine, 1/2" by 7/8"   ribs, 1/2" by 3/4"
keelson, 3/4" by 1 3/4"   floors, 1/2" by 7/8"

paddle, 8' 7" (230.5 cm)

16'5-1/8"

ACKNOWLEDGMENTS I wish to express my sincerest gratitude to all the museum staff who facilitated my studies of their collections. Those associated with my research specific to this paper are George Dyson and the Baidarka Historical Society, Dale Idiens and Briony Crozier (National Museum of Scotland), Ms. C. Hack (De Hidde Nijland Stichting), Corine Bliek (Museon), Cunera Buys and Dorus Jansen (Rijksmuseum voor Volkenkunde), Cees Bakker (Westfries Museum), Graham and Roger Pickles (Whitby Museum), Aileen Nisbet (Hunterian Museum), Adrian Zealand (McManus Galleries), Hans Kleist-Thomassen and Lars Peter Danielsen (Peqatigiiffik Qajaq Nuuk), and Emil Rosing, Claus Andreassen, and the staff of the Greenland National Museum. A special thanks to John and Stella Brand, Kirsten Madsen and Anders Gedionsen, and Duncan Winning, O.B.E., for their generous hospitality during my travels.

REFERENCES

Adney, Edwin T., and Howard I. Chapelle
1964   *The Bark Canoes and Skin Boats of North America*. U.S. National Museum Bulletin 230. Washington, D.C.: Smithsonian Institution.

Birket-Smith, Kaj
1924   Ethnography of the Egedesminde District with Aspects of the General Culture of West Greenland. *Meddelelser om Grønland* 46.

Brand, John
1987   *The Little Kayak Book*, part II. Colchester, Essex: John Brand.

Gilberg, Rolf
1984   Polar Eskimo. In D. J. Damas, ed., *Arctic*, Handbook of North American Indians, vol. 5, pp. 577–594. Washington D.C.: Smithsonian Institution.

Nooter, Gert
1971   Old Kayaks in the Netherlands. *Mededelingen van het Rijksmuseum voor Volkenkunde* (Leiden) 17.
1991   East Greenland Kayaks, pp. 319–347 in E. Arima, et al. eds., *Contributions to Kayak Studies*. Canadian Museum of Civilization Mercury Series, Canadian Ethnology Service Paper 122, Ottawa.

Petersen, H. C.
1986   *Skinboats of Greenland*. Trans. K. M. Gerould. Ships and Boats of the North, vol. 1. Roskilde: National Museum of Denmark, the Museum of Greenland, and the Viking Ship Museum.

Peyrère, Isaac de la
1855   Relation du Groenland, pp. 181–249 in A. White, ed., *A Collection of Documents on Spitzbergen and Greenland*. London: Hakluyt Society.

Rasmussen, Knud
1921   *Greenland by the Polar Sea*. New York: F. A. Stokes.

Souter, William Clark
1934   *The Story of Our Kayak and Some Others*. Aberdeen: University of Aberdeen Press.

Steensby, H. P.
1910   Contributions to the Ethnology and Anthropogeography of the Polar Eskimos. *Meddelelser om Grønland* XXXIV:253–407.
1912   Etnografiske og antropogeografiske rejsestudier i Nord-Grønland, 1909. *Meddelelser om Grønland* 50:133–173.

4 A Seventeenth-Century Kayak and the Swedish Kayak Tradition

HUGH COLLINGS

THE SKOKLOSTER PALACE KAYAK

Skokloster Palace is situated on the northern side of Lake Mälar about 11 miles (18 km) south of Uppsala, Sweden. During the Middle Ages, the parish belonged to a Cistercian nunnery. After the Reformation, it became a Crown estate. The estate was donated to Captain Herman Wrangel in 1611. His son, Count Carl-Gustaf Wrangel, began to build the palace in 1654. The final roofing work was completed in 1668; the palace now contains about 80 rooms.

Wrangel's daughter, Margareta (1643–1702), inherited Skokloster, as none of his sons survived him. Upon her death, the palace passed to her son, Abraham Brahe (1669–1728). The palace remained in the Brahe family until 1967, when it became the property of the Swedish Crown. Today it is open to the public and houses one of the world's finest collections of Baroque art and furniture.

Carl-Gustaf Wrangel (1613–1676) was an ambitious campaigner and administrator of Swedish war efforts and territories in northern Europe. He became a field marshal and a lord admiral as well as marshal of the realm. When he was not in the field, he lived in one of his large castles in Swedish Pomerania (northern Germany) or in the Wrangel Palace on the Isle of Knights, Stockholm. Although Wrangel did not live long at Skokloster, he furnished it for himself and his family.

Wrangel was a man typical of the period. Like other leading warriors, politicians, and landowners, he had a special role to play in Swedish history. He drove the Danes from southern Sweden, took part in the Polish and Danish war of King Karl X Gustav, and promoted Swedish interests in Germany. A man in his social position would be well versed in art, literature, and science. All the latest technical developments of the period were collected in the Wrangel Armoury. These were mostly weapons and firearms but also included scientific instruments and exotic objects, curiosa, from other continents. One of these objects is a magnificent and well-preserved Greenland kayak. Today the kayak is on view in the armoury. Bengt Kylsberg kindly showed me the wonders of Skokloster Palace and allowed me to survey the kayak. Assisting in the measuring were Janne Borgström and Howard Coombs. John Brand from the United Kingdom supplied his knowledge about kayaks and surveying.

Origins of the Skokloster Kayak. The Skokloster kayak was first mentioned in a palace inventory of 1710. It is unmistakably from the west coast of Greenland, given its similarity in form and dimensions to other kayaks documented from that region (Figure 4.1).

There is no mention of how the kayak came to be in the Wrangel Armoury. While Carl-Gustav Wrangel was in Pomerania, he collected books, maps, weapons, and scientific instruments from throughout Europe. Wrangel had agents in European capitals who purchased the objects he desired. Several newspapers were available to him and he could have read about the sale of a kayak. Immediately a messenger could have been dispatched to enable his agent to make the transaction and arrange for the kayak to be shipped to Stockholm. The Dutch whale fishery was active during this period, and even today there are a number of old Greenland kayaks in Dutch museums. In Copenhagen, Ole Worm operated a museum that specialized in Greenlandic curiosa. For a wealthy man, the acquisition of a kayak in 17th-century Europe could not have been too difficult.

Figure 4.1—The Skokloster Palace kayak, Wrangel Armoury.

Dimensions of the Skokloster Kayak

Overall:

length (bow knob missing)	569.5 cm (18 feet, 8 ¼ inches)
width, cockpit front	43.8 cm (17 ¼ inches)
depth sheer amidships	20.8 cm (8 ¼ inches)
depth at 75 cm from bow	21.5 cm (8 ½ inches)
depth at 500 cm from bow	13.6 cm (5 ⅜ inches)

Cockpit:

length overall	39.0 cm (5 ⅜ inches)
width overall	38.5 cm (15 ⅛ inches)
height rim forward	44 mm (1 ¾ inches)
height rim aft	40 mm (1 ⅝ inches)
width rim forward	15 mm (⅝ inch)
width rim aft	23 mm (⅞ inch)

Structural members, at cockpit:

sheer planks	20 x 65 mm (¾ x 2 ½ inches)
keelson (on flat)	20 x 29 mm (¾ x 1 ⅛ inches)
stringers (on flat)	20 x 25 mm (¾ x 1 inch)
ribs	12 x 22.5 mm (½ x ⅞ inch)

Form and Function. Even those unacquainted with kayaks and their function would be struck by the Skokloster kayak's elegance and attractive shape. The gently curved stem and the higher, upswept stern convey a feeling of power, grace, and efficiency. Here are some details:

- The forefoot is deep and the stern is rather shallow.
- Cross sections forward of the cockpit are fuller and deeper than the aft sections, which are shallower and more V bottomed.
- The sides have a fair amount of flair and the floor is quite narrow (11 ¾ inches [30 cm] in the cockpit)
- Both forward and aft decks are flat. The deck directly in front of the cockpit hoop was probably in its original condition, raised slightly; the deck beam supporting the forward end of the hoop is not in place.
- Five deck lines are without bone fittings: three forward and two aft of the cockpit.
- The cockpit hoop is made of a red-colored wood (Figure 4.2). At the front and sides of the hoop, the skin covering is attached with bone pegs set into the inside upper edge of the hoop. At the back of the hoop, the bone pegs are set into the outside edge, which would effectively stop the pegs from digging into the kayaker's back. There is a shallow indentation on the outside surface of the hoop where the join occurs. Indentation and lip would provide a secure hold for the bottom edge of the kayaker's anorak, enabling the kayak to be used in heavy seas and rolled without shipping water.
- There is a bone knob at the stern point; stem knob is missing.
- The keel at the bow is shod with a bone skid that would protect the skin from wear and tear.

Inference of performance is a matter of experience and comparison, intuition, and intelligent guesswork. The kayak's length gave it speed, and in skillful hands, the narrow V bottom made for good rough-water handling and tracking. In a following sea, in which the wind is blowing in the same direction as the kayak is being paddled,

Figure 4.2—Coaming of the Skokloster Palace kayak; port profile showing concave section forward and added lip aft for *tuilik*.

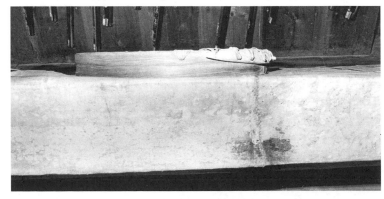

the relative lack of buoyancy in the stern would prevent the bow from plunging too deeply into a wave. The upswept stern would give ample reserve buoyancy to lift the rear deck when submerged.

Very few, if any, of today's reinforced-plastic kayaks, which are modeled on Greenland kayaks, have the same function (or dimensions) as the old Greenland designs. The recreational paddler has other standards and goals when compared with the kayak hunter. The seal-catching characteristics of a kayak are not necessary for the recreational kayaker. Therefore, the only way to really know and feel how this kayak performs is to construct a replica.

Internal Construction. The construction of the kayak is not unlike the method used to build model airplanes before the introduction of plastics. It is basically a wooden framework covered with sealskin. For a hunting community well acquainted with zoology, it would have been natural to copy the construction elements found in sea mammals. It takes skill to make a well-designed, symmetrical kayak using relatively few tools.

The joinery shows little sign of having been worked with European tools. The stringers look as though they have been split and then roughly squared off. Both sheer planks have been joined in the cockpit area using hook scarfs. The joint has been held together with skin ties, the holes for which have been drilled along the join line (Figure 4.3).

The ribs have been split from small branches, the inside edge of each rib retaining its natural rounded shape. The ribs have been

Figure 4.3—Cockpit from starboard with loose front and rear deck beams.

tenoned into the bottom edge of each sheer plank. The mortise holes are narrower than the ribs; the ends of the ribs have been shouldered to fit into the holes.

The ends of the deck beams have been chamfered to fit into mortise holes along the inside faces of the sheer planks. The mortise holes are blind. In order to hold the bottom edges of the sheer planks at the correct angle and to provide a measure of rigidity, in a few places the bottom edges have been tied together using straps across the kayak.

The forward deck beam or *masik*, which supports the hoop, is no longer in place but is attached to the kayak with a thong. The under edge of the *masik* has a short projection that worked as a hand hold, permitting the kayaker to pull his legs and trunk into the tight-fitting cockpit area. The wedging of the kayaker in the kayak enabled him to perform advanced paddle strokes and self-rescue techniques that greatly enhanced the chance of survival in harsh conditions.

The Paddle. The paddle is double-bladed with long narrow blades in the same plane. (Recreational paddlers often use feathered paddles with the blades set at 75 to 90 degrees to each other.) One of the bone end fittings is missing. The fitting is attached to the end of the blade with a mortise-and-tenon joint, the mortise being in the fitting. The blades are edged with bone strips attached with small bone pegs. The paddle has been repaired in several places using thong bindings and dowels. Despite that, it is of normal length, 6 feet, 10½ inches (209.4 cm). Blade width is 3¼ inches (8.1 cm).

Developmental Aspects. The Skokloster kayak (Figures 4.1, 4.4) is an early example of the deep V-bottom design, which is commonly believed to be typically Greenlandic but actually developed in the 17th century, apparently on the central west coast. The older, shallow V bottom, natively considered flat, continues in South and East Greenland to the present. Although 17th-century examples comparable to Skokloster survive elsewhere, notably in Holland, lines drawings of them are currently unavailable. The drawing of the Skokloster example should enable researchers to compare it more closely with other designs. The shape demonstrates strong continuity with the earlier flat-bottom Greenland design as represented by the 1606 Lübeck kayak and, by extension, with the East Canadian Arctic forms. Most salient in the linkage is the deep bow and shallow stern configuration, including the profile shapes, especially in the bow. Internal construction remains in the old manner without the variously fitted-together cutwater boards at the ends. Deck

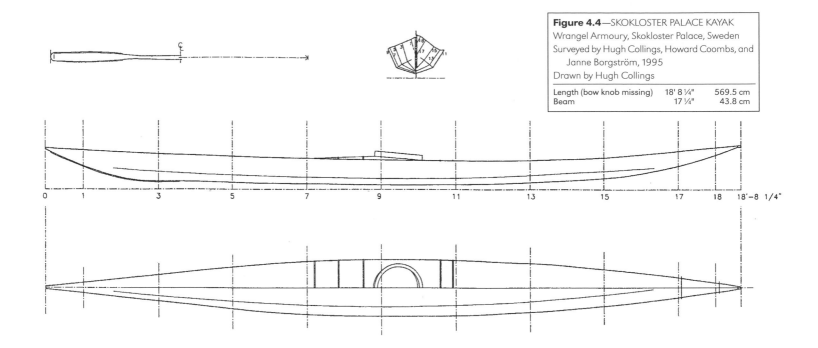

Figure 4.4—SKOKLOSTER PALACE KAYAK
Wrangel Armoury, Skokloster Palace, Sweden
Surveyed by Hugh Collings, Howard Coombs, and
Janne Borgström, 1995
Drawn by Hugh Collings

| Length (bow knob missing) | 18' 8 ¼" | 569.5 cm |
| Beam | 17 ¼" | 43.8 cm |

beams are not yet tenoned through the sheer boards but rest in comparatively shallow notches, the whole drawn together by lashings as may be seen in the accompanying interior detail photo (Figure 4.5). Skokloster is one of a handful of surviving 17th-century examples that show that the **V** bottom of West Greenland design developed first, before the assorted end shapes, sheer, and modern framework construction. The Skokloster kayak has lasted three centuries in remarkably good condition, a jewel of master kayak design from more intense sea-hunting days.

At the other end of historical development are the next two models to be described. They have been adapted to accommodate the size and height of Europeans. The first, a West Greenland built on Disko Island (Map 2) in 1916 for Thorild Wulff, is oversized for the hefty botanist but otherwise looks normal in shape. The second, built in East Greenland in 1968 for writer Jörn Riel, remains in the usual range of dimensions but has a very noticeably bulged-up foredeck to accommodate the feet. The shallow **V**-bottom East Greenland design may be less amenable to simple enlargement overall than the deeper **V** West Greenland because of volume considerations. The builder was probably more concerned with performance characteristics such as speed, stability, tracking, turning, wind, and wave response than with aesthetics. A little background to the study of these two Europeanized models follows.

Figure 4.5—Interior aft showing old deck beam lashing method.

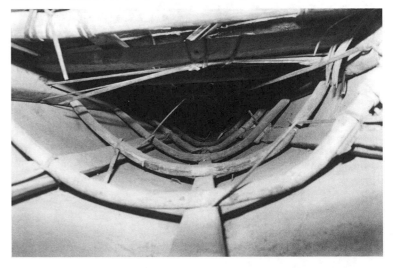

THE SWEDISH KAYAK TRADITION AND THE "SWEDE" FORM

In the early 1980s, just before moving from England to Sweden, I had the good fortune to meet John Brand of Colchester. He had already spent a lot of time surveying and documenting arctic kayaks in British and Danish museums. He showed me his immaculate draftsmanship, fired my enthusiasm, and initiated an understanding of good design: the balance between form and function. In 1988, I contacted Christian Lagerkrantz of the Folkens Museum, Stockholm, and obtained permission to survey a couple of kayaks. The museum's boat storage was packed with a worldwide collection of canoes and kayaks, including a rare Maritime Chukchi kayak collected by A. E. Nordenskiöld. The two Greenland kayaks built for Europeans were chosen and surveyed during two intensive days with the able assistance of Howard Coombs, an authority on small craft.

One motive for carrying out the surveys was to enrich and inspire the already strong Swedish kayak tradition. In 1869, John McGregor toured Sweden in one of his Rob Roy clinker-built canoes. (In Sweden, as in the United Kingdom, the terms *canoe* and *kayak* are interchangeable; a "Canadian canoe" is an open canoe.) His journey became a sensation. In 1873, Commodore Carl Smith, the father of Swedish canoeing, purchased an English canoe while on naval service in Malta and brought it back to Sweden. Soon others were imported from England, and local boat builders began to produce a variety of paddling (double-paddle only) and sailing canoes. These were exclusive and expensive. Around the turn of the century, canoeists began to build more modest canoes using canvas-on-wooden-frame construction. Magazines printed drawings for the home builder and designs steadily improved. Paddler/designers like Gerhard Högborg and Sven Thorell were especially prolific.

Slender racing canoes already had dimensions and forms similar to their counterparts of today. Designers copied the sharp "double-wedge" shape of contemporary low-resistance motorboats, which were deep at the stem with the hull becoming flat and shallow at the stern. These long, narrow boats were considered the ideal shape for racing or fast touring kayaks. The idea was that the sharp bow would cut through the water, which would exit easily under the flat stern. Although these canoes were fast, they became difficult to manage in a following sea when the double-wedge or "swede" form was extreme. The term *swede form* is in use today, denoting a canoe that has its widest beam aft of the midpoint. Some East Canadian Arctic and Polar Eskimo kayaks once had such a form.

In 1919, Sven Thorell made drawings for a *långfärdskanot* (traveling kayak), about 16 feet (5 m) long and 26 inches (66 cm) wide. The kayak, called "Åland" after a group of islands between Sweden and Finland, had a very slight swede form. The design has proved to be exceptionally seaworthy and is still manufactured today in reinforced plastic by Ingvar Ankervik of Vituddens Kanotvarv (VKV), a canoe workshop that has been in business since the early 1920s.

The home-build tradition is strong in Sweden, no doubt due to the availability of canoe plans and good-quality timber. Between the two world wars, many thousands of boys and young men, often with very meager resources (jute sacks replaced the more expensive canvas) were putting together their own canoes in cellars or attics. I hope that the availability of drawings of these two Greenland designs will encourage a continuation of home-built craft tradition in the arctic kayak style.

THORILD WULFF'S WEST GREENLAND KAYAK

This kayak (Figure 4.6) was collected by the Swede Thorild Wulff during Knud Rasmussen's Second Thule Expedition in 1916. The expedition's goal was exploration of the most northerly part of North Greenland. Wulff was a renowned botanist and ethnographer who had traveled to Iceland, Siberia, and the Far East. During the coastal journey north by boat, Wulff visited several West Greenland settlements. He observed the hunters in their kayaks and was determined to practice *kajakrodd* (kayak rowing). In Aasiaat (Egedesminde) (Map 4), he tried in vain to squeeze himself into a kayak.

A specially built kayak was constructed for Wulff on Disko Island, but when he tried it out he discovered that there was not enough room for his legs, and so the kayak had to be rebuilt. In his *Grönländska Dagböcker*, or *Greenland Diaries*, there is a photograph with the caption "Wulff practices kayak rowing at Disko" (Wulff 1934:81). The kayak is unmistakably the one in the Folkens Museum. Wulff sits there rather stiffly with the paddle at rest, one tip of the blade touching the calm and almost ripple-free water. He is warmly dressed in a haired sealskin anorak. On one side of the kayak, floating on the water just aft of the cockpit, is a long narrow object. It may be a sealskin pontoon once used by young boys when learning to balance. Before Wulff continued north, the kayak, together with the paddle and other ethnographic objects, was crated for shipment to Stockholm.

The Second Thule Expedition ended in a double tragedy. First, the Greenlandic explorer Olsen became lost and, despite a search, was never found. The party that Wulff belonged to later crossed an area devoid of game. The members became hungry and extremely weak. Wulff became ill and eventually so enfeebled that he was unable to continue. He begged his companions to leave him and try to save themselves. This they did. They returned later to rescue him, but he was never found.

Dimensions

Overall:

length (stern knob missing)	534.0 cm (17 feet, 6 ¼ inches)
width, cockpit front	57.0 cm (22 ½ inches)
depth sheer, amidships	22.1 cm (8 ¾ inches)
depth 100 cm from bow	25.6 cm (10 inches)
depth 480 cm from bow	17.4 cm (6 ⅞ inches)

Cockpit:

length overall	46.2 cm (18 ⅛ inches)
height rim	36 mm (1 ½ inches)

Structural members, at cockpit:

sheer planks	18 x 78 mm (¾ x 3 inches)
keelson & stringers (on edge)	19 x 27 mm (¾ x 1 inch)
ribs	10 x 24 mm (⅜ x 1 inch)

Form and Function. By West Greenland standards, this is a large kayak. The builder must have made allowance for Wulff's fairly stout figure and, more important, appreciated the difficulties Europeans have in bending their legs the "wrong" way. Young Greenlanders are taught at an early age how to exercise their knee joints, which enables them to squeeze into their low-decked craft. Despite the kayak's size, it has pleasing lines. The chine stringers are placed well out from the keelson, making a deep V hull that provides seaworthiness and directional stability. In order to keep the kayak from lying too high in the water, it has been given a fair amount of rocker in the keel line. The bone keel skid strip at the stern, 25 ⅜ inches (64.5 cm) from the tip, has been drilled with two holes, center-to-center distance 11 ⅛ inches (28 cm), for attaching a wooden skeg that would counteract any tendency of the kayak to the left or right after each paddle stroke.

In a following sea, the relative lack of buoyancy in the shallow stern compared with the bow would prevent the bow from plunging too deeply into a wave. The upswept stern would give ample reserve buoyancy to keep the rear deck free from water. This is undoubtedly a stable and seaworthy design. The larger than usual dimensions would give ample leg room and stowage for camping gear for those considering replication. The kayak is in good condition and the framework joinery of a high standard.

The paddle is double-bladed with long, narrow blades in the same plane. The bone end fitting is attached to the blade with a mortise-and-tenon joint, the mortise being in the fitting. The blades are edged with bone strips attached with small bone pegs.

JÖRN RIEL'S EAST GREENLAND KAYAK

This relatively short kayak from East Greenland (Figure 4.7), built in Inigssalik by Masanti, was used by the Danish writer and ethnographer Jörn Riel for a 186-mile (300 km) journey along the East Greenland coast in July 1968. The kayak was donated to the Folkens Museum by Riel, who was married to a Swede and lived in Sweden for a short time. He has written many novels and children's books about Greenland and Greenlanders.

Dimensions

Overall:

length	500 cm (16 feet, 4 ⅞ inches)
width, cockpit front	50.8 cm (20 inches)
depth sheer amidships	17.0 cm (6 ¾ inches)
depth 100 cm from bow	18.5 cm (7 ¼ inches)
depth 400 cm from bow	17.8 cm (7 inches)

Cockpit:

length overall	45.0 cm (17 ¾ inches)
width overall	42.0 cm (16 ½ inches)
height rim	36 mm (1 ½ inches)
width rim	24 mm (2 inches)

Structural members, at cockpit:

sheer planks	14 x 51 mm (½ x 2 inches)
keelson & stringers (on edge)	18 x 29 mm (¾ x 1 ⅛ inches)
ribs	19 x 31 mm (¾ x 1 ¼ inches)
front deck beam	15 x 45 mm (5/8 x 1 ¾ inches)
rear deck beam	19 x 50 mm (¾ x 3 inches)
floor boards	9 x 75 mm (⅜ x 3 inches)

Form and Function. This kayak has the "normal" East Greenland shape—almost flat-bottomed and shallow hull, steeply angled sides, long overhang at ends, and flat deck. The two exceptions are the relatively short length (usual length is about 18 feet [550 cm]) and

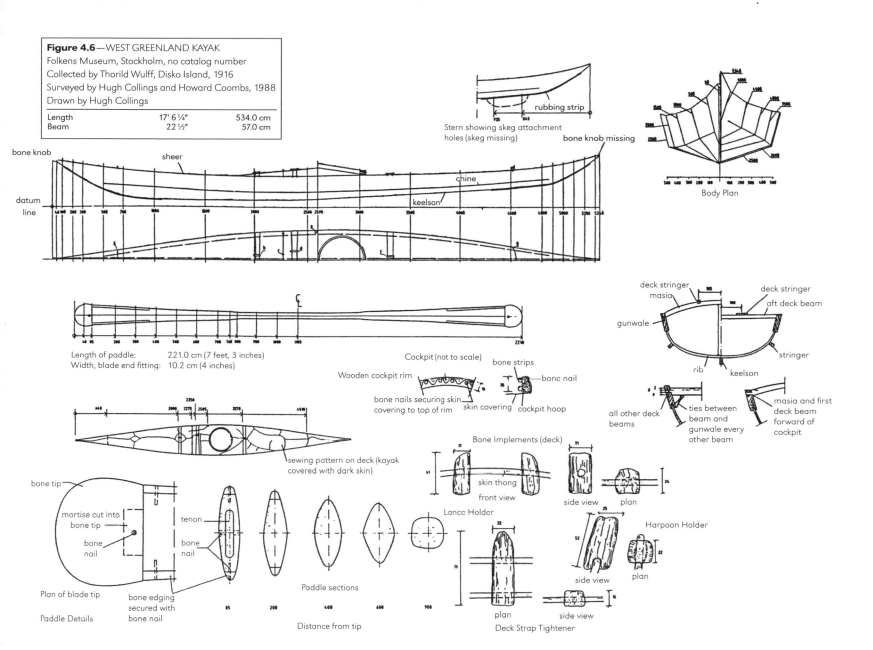

Figure 4.6—WEST GREENLAND KAYAK
Folkens Museum, Stockholm, no catalog number
Collected by Thorild Wulff, Disko Island, 1916
Surveyed by Hugh Collings and Howard Coombs, 1988
Drawn by Hugh Collings

| Length | 17' 6 ¼" | 534.0 cm |
| Beam | 22 ½" | 57.0 cm |

bone knob

sheer

datum line

chine

keelson

Stern showing skeg attachment holes (skeg missing)

rubbing strip

bone knob missing

Body Plan

Length of paddle: 221.0 cm (7 feet, 3 inches)
Width, blade end fitting: 10.2 cm (4 inches)

Cockpit (not to scale)

bone strips

Wooden cockpit rim

bone nail

bone nails securing skin covering to top of rim

skin covering

cockpit hoop

deck stringer masia

deck stringer

aft deck beam

gunwale

rib

keelson

stringer

all other deck beams

ties between beam and gunwale every other beam

masia and first deck beam forward of cockpit

sewing pattern on deck (kayak covered with dark skin)

Bone Implements (deck)

skin thong

front view

side view

plan

Lance Holder

Harpoon Holder

bone tip

mortise cut into bone tip

tenon

bone nail

bone nail

Paddle sections

plan

side view

Plan of blade tip

Paddle Details

bone edging secured with bone nail

85   200   400   600   900

Distance from tip

plan

side view

Deck Strap Tightener

the greater depth in front of the cockpit. The short length might allow easier sledge transportation during the winter. The short waterline length would limit top speed; however, I have paddled a number of short kayaks with relatively low wetted surface area and have found them to be surprisingly fast. My experience in the East Greenland type is very limited and it is difficult to have an opinion about the seaworthiness of this kayak.

The kayak has both original paddle and paddle holder. The latter is a short piece of shaped wood that is held in place by the deck straps in front of the cockpit. One end of the paddle is placed under the paddle holder and the rest lies perpendicular to the side of the kayak with the blade on or just below the surface of the water to act as a stabilizer.

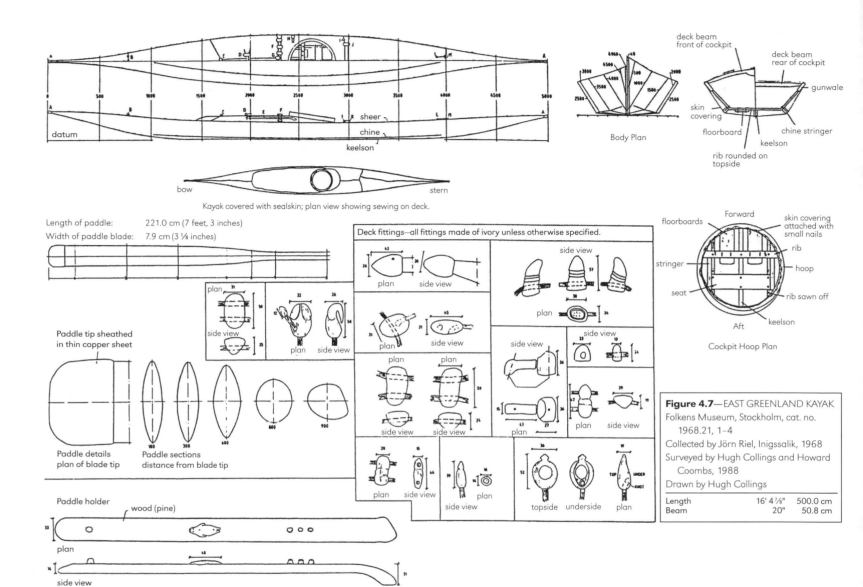

Length of paddle: 221.0 cm (7 feet, 3 inches)
Width of paddle blade: 7.9 cm (3 ⅛ inches)

Paddle tip sheathed in thin copper sheet

Paddle details plan of blade tip

Paddle sections distance from blade tip

Paddle holder
wood (pine)

plan

side view

Kayak covered with sealskin; plan view showing sewing on deck.

bow   stern

datum   sheer   chine   keelson

Body Plan

deck beam front of cockpit
deck beam rear of cockpit
gunwale
skin covering
floorboard
chine stringer
keelson
rib rounded on topside

Forward
floorboards
skin covering attached with small nails
stringer
rib
hoop
seat
rib sawn off
keelson
Aft

Cockpit Hoop Plan

Deck fittings—all fittings made of ivory unless otherwise specified.

plan   side view
side view
plan   side view
plan   side view
side view
side view
plan   plan
side view   side view
plan
side view
plan   side view
plan
plan   side view
plan
side view   plan
topside   underside   plan

Figure 4.7—EAST GREENLAND KAYAK
Folkens Museum, Stockholm, cat. no.
  1968.21, 1–4
Collected by Jörn Riel, Inigssalik, 1968
Surveyed by Hugh Collings and Howard
  Coombs, 1988
Drawn by Hugh Collings

| Length | 16' 4⅞" | 500.0 cm |
| Beam | 20" | 50.8 cm |

The kayak is in fair condition and the framework joinery is of a simple standard. The keelson and chine stringers are attached to the ribs with galvanized nails—a common practice on the east coast. There are quite a few gaps between the ribs and stringers that have been filled out with slim wooden blocks. Many of the bone fittings are of a high artistic standard; I wonder if they are older than the kayak. Some of the fittings for the cross straps are phallic in shape—perhaps visual puns. A superbly carved seal ornaments the paddle keeper.

REFERENCE

Wulff, Thorild
1934 Thorild Wulffs *Grönländska Dagböcker*. Edited by Axel Elvin. Stockholm: Albert Bonnier.

5

Kayaks in England, Wales, and Denmark:
Excerpts from *The Little Kayak Book* Series

JOHN BRAND
Edited by E. Arima

There are many kayak specimens with Greenlandic designs vital for study that are curated in northern European museums. John Brand of Colchester, Essex, an architect, is the premier kayak researcher in the U.K. From 1984 through 1988, he published The Little Kayak Book *series in three charming volumes full of fine drawings. He has generously given permission to publish a selection of his drawings and commentary on nine Greenlandic and two East Canadian kayaks. Brand has devoted special attention to their individuality, an aspect too often neglected. He provides extensive details on measurements, provenance, survey circumstances, and replication possibilities. Brand's thoughts on performance draw on long experience sea racing and building Greenlandic and Aleut "semi-replicas."*

Metric measurements have been retained in the lines drawings. The designation "bKr" refers to the specimen numbers assigned by John Brand and stands for "Brand Kayak research." Text has occasionally been omitted, so citations to The Little Kayak *series are provided at the end of each paragraph. Spelling and punctuation have been edited for clarity. —Ed.*

THE 1613 KAYAK FROM SOUTHWEST GREENLAND (bKr–001)

Trinity House, Hull, Humberside, England

length	361.3 cm (11 feet, 10 ¼ inches)
width	43.0 cm (17 inches)

It is appropriate to give this kayak [Figure 5.1a] the first kayak record number [bKr–001] because it is the oldest known to me being approximately a hundred years older than the better known specimen at Marishal College, Aberdeen, Scotland. In the 1960s the Hull kayak was rumoured to have a bone frame which made the idea of a survey doubly attractive (1984:3).

The impetus for this survey came from the other side of the Atlantic: John D. Heath's long-standing interest in the ethnological aspects of the subject naturally focussed on this kayak. It may have been seen as an evolutionary pivot from which some eastern kayaks could be traced and earlier forms inferred from the pattern of development (1984:3).

The survey did not produce any proof [of] the existence of a bone frame. We did not know until we arrived at Trinity House that a determined character had established squatter's rights in the kayak and would prevent us sending a positive report back across the Atlantic: the [effigy] had been so well lashed in 347 years earlier, and all the skins so thickly coated over, that the idea of inspecting the interior vanished immediately (1984:3).

Considering its great age, we thought that the frame had survived the centuries very well, being small and belonging to Trinity House were the great advantages no doubt. The major area of collapse was abaft the cockpit… (1984:5).

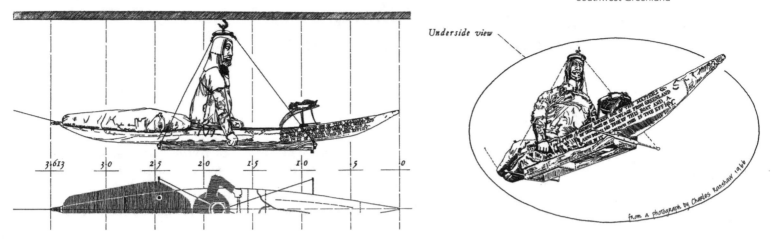

Figure 5.1a
THE 1613 KAYAK (bKr–001)
Southwest Greenland

Underside view

3 613   3 0   2 5   2 0   1 5   1 0   5   0

from a photograph by Charles Ranshaw 1966

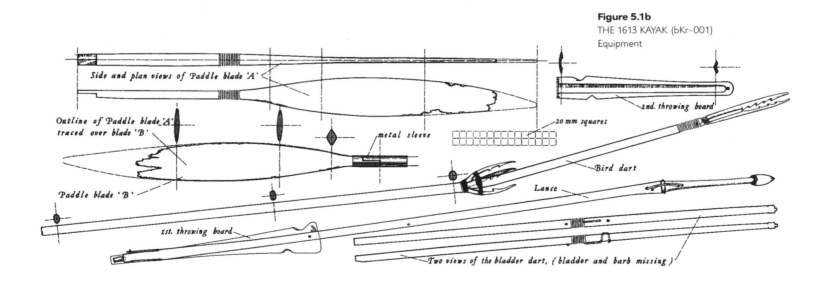

Figure 5.1b
THE 1613 KAYAK (bKr–001)
Equipment

Side and plan views of Paddle blade 'A'

Outline of Paddle blade 'A'
traced over blade 'B'

metal sleeve

2nd. throwing board

20 mm squares

Paddle blade 'B'

Bird dart

Lance

1st. throwing board

Two views of the bladder dart, (bladder and barb missing)

The float on the aft deck is most remarkable: before being stuffed in Hull it was blown up through two symmetrically placed tubes with stoppers. It needed to be part of the boat if there was a capsize but there were no clues about the way it was secured. The streamlined shape is interesting; I feel that its purpose was to make kayaking safer for someone who was really too large for a small boat. The foreparts of the float were splayed back for ease of paddling and the overhangs at the stern showed the need to provide more buoyancy there. These features and its great length relative to the length of the kayak make the float unique in my experience (1984:5–6). [A large float could also be useful for hunting large seals. —Ed.]

I did not start to evaluate the shape of the paddle blades [Figure 5.1b] until I redrew them in March 1983. I traced what I considered to be the original shape of one blade and placed the tracing paper over the second blade. This showed that each was accurately made and then I realized that I was looking at leaf-shaped blades. Of course, the fine tips that came automatically with the use of the spline are a matter for discussion, but it would be most unusual to have graceful sides ending in a blunt edge (1984:6).

The most obvious point about the rest of the equipment [Figure 5.1b] is that the tray for the harpoon line is roughly made compared to the weapons. In his letter (15 May 1966), John Heath observed: "The hunting implements are interesting. It looks as if the lance has been made from an old winged harpoon or *ernangnak* as they call it, by adding a different foreshaft and head. The second throwing stick could be used for both the other missiles shown, the bladder dart (with bladder and barb missing) and the bird dart. The latter is interesting because of its multi-pronged head; 'more recent' bird darts have a single point at the head, but the side prongs remained the same, varying slightly among different specimens." The elegance of the weapons is easy to appreciate: they look sporty but I think this comes from wanting the lightest type of missile for a teenager under instruction (1984:8).

The measurement from the gunwale to the effigy's forehead was 860 mm [33 feet, ⅞ inches], about 100 mm [four inches] more than a tall man.... This may be why the display looks a little surreal, that and the [fisheye] view.... The anorak was stretched tightly from the shoulders to the deck, which indicates that every effort was made to heighten the dramatic effect (1984:8).

In my letter to John Heath, 1st February, 1966, I made my first guesses: "Charlie [Ranshaw] and I are both strongly of the opinion that the kayak was from Greenland.... [I]t may have been made for a boy and the boy may still have been using it when he was 16, 17, 18 or so, when he had really out-grown it." As soon as he received Charles Ranshaw's photographs, [Heath] replied [in May]: "I concur with you and Charles Ranshaw that it is from Greenland and that it is a boy's kayak. Its external appearance is that of a circa twentieth-century kayak from the Central West Greenland coast, say, around Godthaab" (1984:9).

BAFFIN ISLAND KAYAK (bKr–004)

University of Cambridge Museum of Archaeology and Anthropology, Cambridge (no. 46.482)

length	631.0 cm (20 feet, 8 ½ inches)
width	71.6 cm (28 ⅛ inches) at 362.0 cm station
depth sheer	21.8 cm (8 ½) at 337.0 cm station
paddle length	284.0 cm (9 feet, 3 ¾ inches)
blade width	75 mm (3 inches)

Collected 1934 from Clyde Inlet, Baffinland, by T. T. Paterson, Wordie Arctic Expedition. Surveyed 28 June 1975, Charles Ranshaw, Tony Jefford, Michael O'Connel, John Brand; 13 September 1975, Charles Ranshaw, John Brand; 11 October 1975, Charles Ranshaw, David Zimmerly, John Brand; 4 November 1976, Stella and John Brand (Brand 1984:22).

Many people interested in semi-replicas start with a Baffin Island type [Figure 5.2a, b] and for a heavy kayak they are rewarding craft. I thought there were two main drawbacks in addition to the weight: the extra length makes for difficulties when transporting it to water and the flat bottom gives too much initial stability for use in the sea: thus the Clyde Inlet kayak [multichine] should be a better boat for use in the U.K. (1984:28).

Figure 5.2a
BAFFIN ISLAND KAYAK (bKr–004)

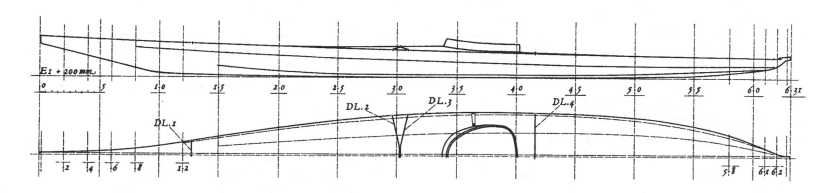

Figure 5.2b
BAFFIN ISLAND KAYAK (bKr–004)
Cross-sections and coaming

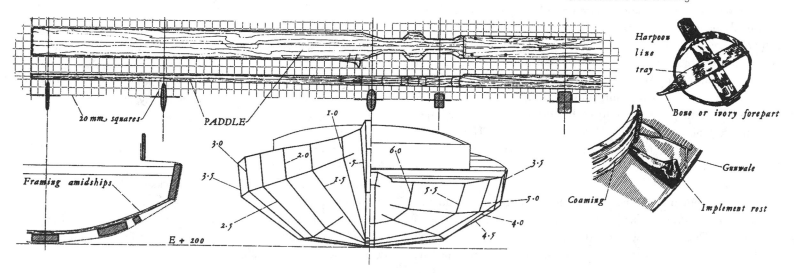

WEST GREENLAND KAYAK (bKr–6)

British Museum, Ethnographical Department (1953-AM.11)
(formerly Bethnal Green Museum, no. 146)

length	585.5 cm (19 feet, ⅝ inches)
width	42.2 cm (16 ⅝ inches) at 276.5 cm station
depth sheer	16.1 cm (6 ⅜ inches) at 250.0 cm station
paddle length	208.6 cm (6 feet, 10 ⅛ inches)
blade width	73 cm (2 ⅞ inches) near tip

Surveyed 1963, Charles Ranshaw, Brian Skilling, John Brand.

Of all the kayaks I have seen, AM.11 [Figure 5.3] comes nearest to the conventional European view of "the" Eskimo kayak. It is long, slim, and unstable and it is from Greenland; it would have lain low in the water and it is certainly pleasant to imagine it being propelled swiftly between menacing ice floes, the sleek shape silently creasing a mirror-like surface.... Only the smallest Caucasians are likely to be able to enter a replica of AM.11 and then they face the problem of not having enough strength to use it seriously: it seems safe to assume that the original owner was an exceptional man since it has an air of extreme elegance that a committee of experts could not bestow; either way such a good roller would challenge other paddlers (1987:8).

The kayak from the Bethnal Green Museum is one which would be worth recreating simply for the joy of seeing its fine lines. Even if one had no interest in kayaking, its value as a piece of sculpture would appeal to many people (1987:10).

Judging by its proportions it is likely that AM.11 was intended for use in fjords and if a working replica can be considered seriously then its normal use ought not to extend beyond touring on calm waters (1987:10).

Quintin Riley was one of the few Englishmen to use an unmodified kayak in Greenland and he assured me in 1964 that early training allowed Eskimos to bend their legs the wrong way a little and having shorter legs and smaller kneecaps must help getting into kayaks as well, which means that the authentic proportions of traditional kayaks soon disappear if the finely judged foredecks are raised. It is equally interesting to observe that kayaks made by Greenlanders for white men have not been as successful as one might have expected (1987:10).

Figure 5.3
WEST GREENLAND KAYAK (bKr–6)

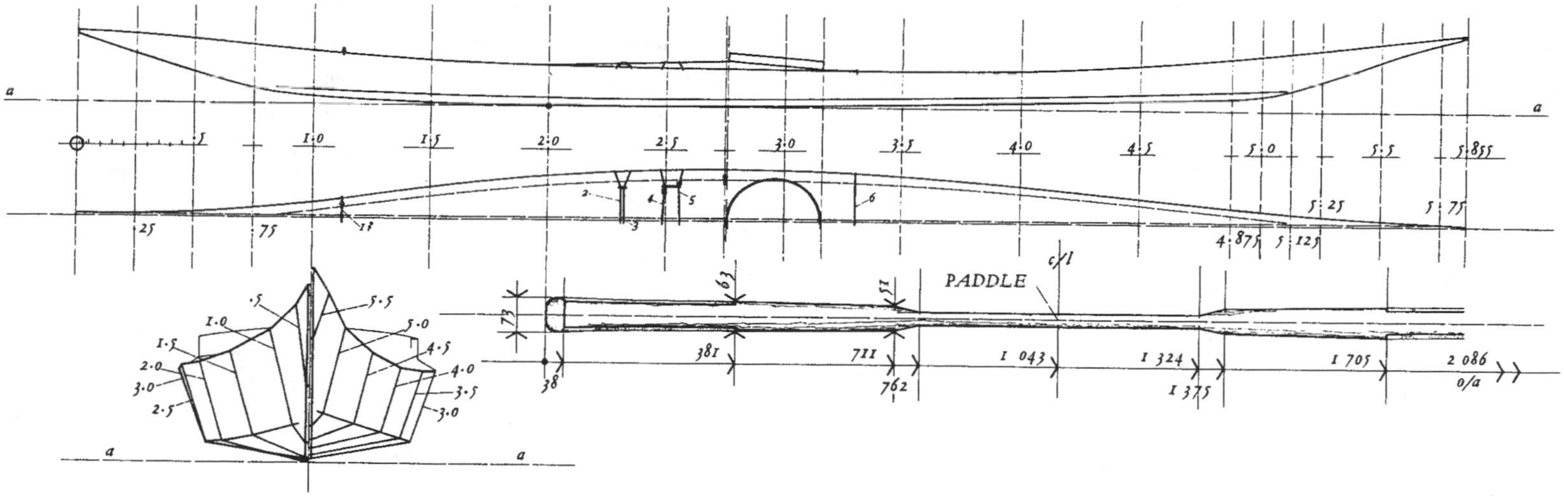

WEST GREENLAND KAYAK (bKr–9)

University of Cambridge Museum of Archaeology and
Anthropology, Cambridge (Z.15360)

length	499.1 cm (16 feet, 4 ½ inches)
width	51.1 cm (20 ⅛ inches) at 243.8 cm station
depth sheer	18.9 cm (7 ⁷⁄₁₆ inches) at 243.8 cm station
paddle length	209.1 cm (6 feet, 10 ⅜ inches)
blade width	7.8 cm (3 inches)

Surveyed 1964, Charles Ranshaw and John Brand.

The idea [that this kayak] [Figure 5.4a, b, c] [is] suitable for recreational kayaking still seems to be a good one despite me going about the job the wrong way in 1978: to save time and money I plunged into making a plug, moulds, and so on without bothering with a conventionally built test boat and launched a GRP [glass resin plastic] kayak in September 1979 to find that a year's spare time working had been wasted (1987:30).

The first reason for this was a mistake made at the working drawing stage when it was noticed that the aft chines were pinched in.... The pinching in was faired out in a happy state of mind because it was believed that the original builder had wanted this and somehow failed; in addition I thought I could not be going wrong technically because more buoyant lines would make the boat safer and easier to handle. Of course, the exact reverse was true—the more buoyant lines lifted the stern to the point where it had very little grip in the water, the skeg at the aft foot was very shallow anyway and perhaps the kayak was only used in Greenland waters when it was well-loaded, but certainly the semi-replica was skittish in small waves and when I had to crash over the wake of a speed-boat in the estuary, the lack of control was more frightening than exciting (1987:30).

The second reason was a mistake that was even more fundamental and naturally it reinforced the first failure. Because my weight exceeds the average for Englishmen and Greenlanders I thought it reasonable to increase all the survey dimensions by one-tenth. This proved to be much too much.... Not only did my GRP [glass resin plastic] version feel too large but I had no idea at the time how delicate was the balance between a kayak for a small man and one for a large man. I repeated these mistakes when making two versions of the Aleut *baidarka* in the Lowie Museum [now the Phoebe A. Hearst Museum of Anthropology] in the United States. Even [on] the second attempt increases [in size] produced unexpectedly large effects, and even with a beam of 53 cm (20⅞ inches) the replica *baidarka* lost some of its liveliness. Even + ¹⁄₄₀ for the lengths and + ¹⁄₂₀ for the breadths was too much. I made no increases for the depths although I now know that, in theory at least, I ought to have decreased the depths (1987:30, 32).

[T]he [University of Cambridge Museum] replica would have suited me if I had altered the lengths by + ¹⁄₂₁, breadths by + ¹⁄₃₄ and the depths by – ¹⁄₁₃, the last being necessary to offset the gains in volume given by the first two. [A]t last it was beginning to dawn on me that displacement was a major obstacle to merely sizing up boats to make them fit the larger sized canoeist (1987:32).

An actual size version of Z.15360 would be a good kayak in calm conditions in the U.K. It should suit many teenagers...(1987:32).

Figure 5.4a
WEST GREENLAND KAYAK (bKr–9)

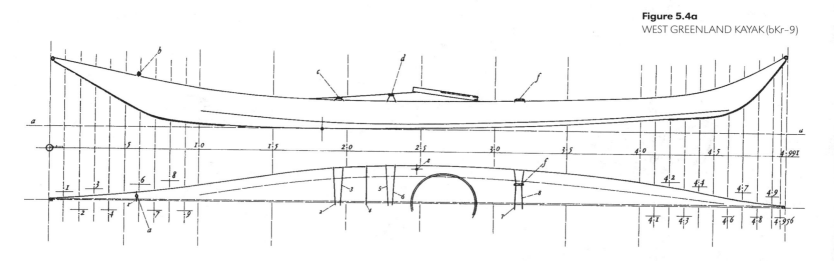

Figure 5.4b
WEST GREENLAND KAYAK (bKr–9)
Cross-sections and coaming

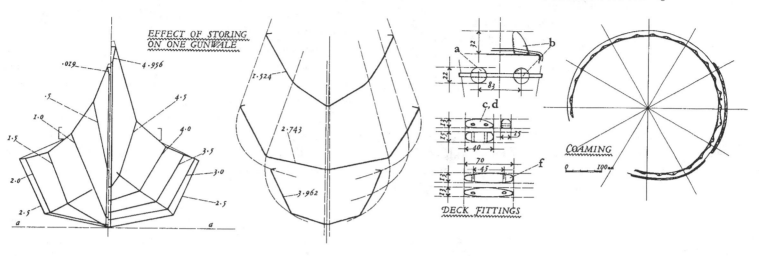

EFFECT OF STORING
ON ONE GUNWALE

COAMING

DECK FITTINGS

Figure 5.4c
WEST GREENLAND KAYAK (bKr–9)
Equipment

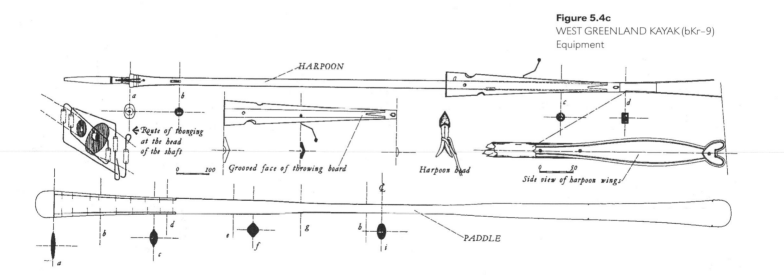

HARPOON

Route of thonging
at the head
of the shaft

Grooved face of throwing board

Harpoon head

Side view of harpoon wings

PADDLE

LABRADOR KAYAK (bKr–10)

Danish National Museum, Ethnographical Department,
Copenhagen (Pb. 1, DK 1471)

length	618.0 cm (20 feet, 3 ⅓ inches)
width	62.8 cm (24 ¾ inches) at 400.0 cm station
depth sheer	23.4 cm (9 ³⁄₁₆ inches) at 400.0 cm station
paddle length	192.0 cm (9 feet, 7 inches)
blade width	5.5 cm (2 ³⁄₁₆ inches)

Surveyed 1975, Stella and John Brand.

Many East Canadian craft give the impression that they lack finesse: that is not true of [the Danish National Museum kayak]. The Baffin Island [kayak] [Figure 5.2a, b] is interesting because of its multi-chine hull and careful construction but with Pb.1 [Figure 5.5a] one has the impression [of] a luxury version.... A great deal of extra care seems to have been taken over its shape and finish: in particular, the cockpit coaming [Figure 5.5b] is as solidly considered as the best of modern Danish furniture; the harpoon line tray [Figure 5.5c] is not well finished but the carved legs have an admirable unity, incorporating weapon rests as they do (1987:34).

Comparisons with the Baffin Island kayak [Figure 5.2a] are inevitable: Pb.1 is neater and has stronger, more attractive lines, but the flat bottom, deep forefoot, and shallow after body make bKr–4 look much more evolved and more suited to rough sea conditions; at best, Pb.1 is an estuary boat (1987:39).

When preparing working drawings for a replica it will be important to check my decision to raise the ends. This can be done by replotting the survey dimensions (1987:39).

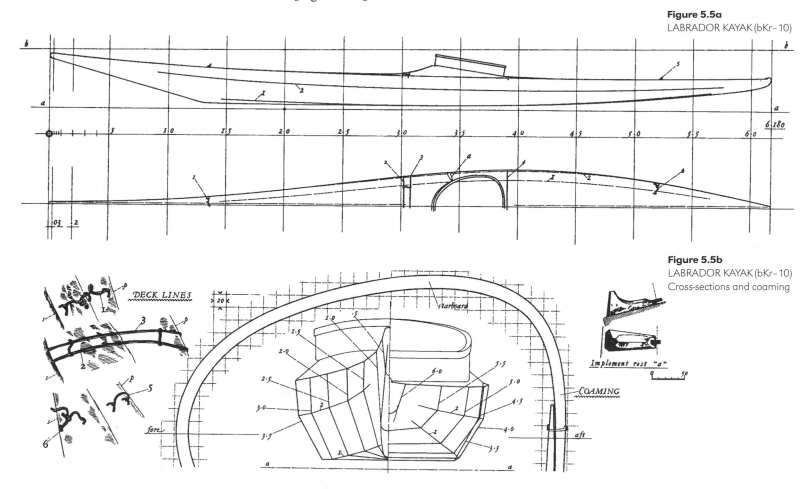

Figure 5.5a
LABRADOR KAYAK (bKr–10)

Figure 5.5b
LABRADOR KAYAK (bKr–10)
Cross-sections and coaming

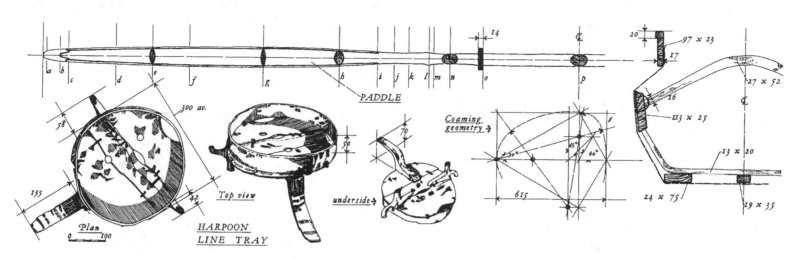

Figure 5.5c
LABRADOR KAYAK (bKr–10)
Equipment and amidships structure

PADDLE

300 av.

58

155

0 Plan 100

Top view

42

70

underside →

HARPOON
LINE TRAY

Coaming
geometry →

30°   45°   60°

615

14

℄

20

97 × 23

17

16

113 × 25

27 × 52

℄

13 × 20

24 × 75

19 × 35

Figure 5.6
EAST GREENLAND KAYAK (bKr–12)
Quintin Riley's Kayak

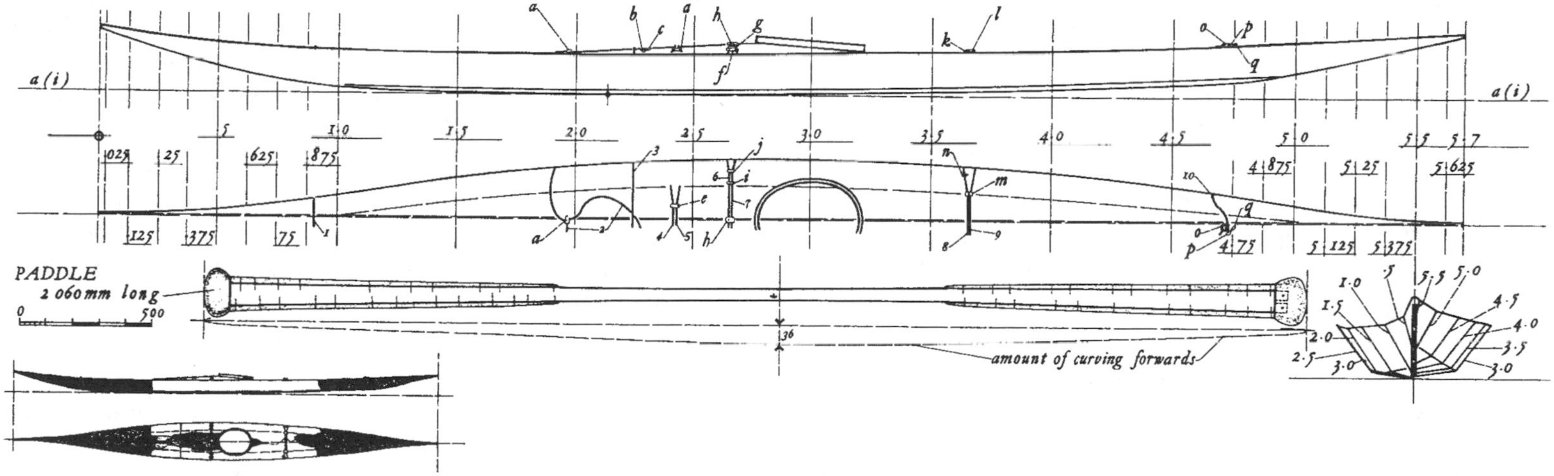

a(i)

a(i)

1025   25   625   875   1·0   1·5   2·0   2·5   3·0   3·5   4·0   4·5   4·875   5·0   5·25   5·5 5·7

125   375   75   5·125   5·375   5·625

PADDLE
2 060 mm long
0   500

36

amount of curving forwards

1·0   ·5   5·5 5·0
1·5   4·5
2·0   4·0
2·5   3·5
3·0   3·0

EAST GREENLAND KAYAK: QUINTIN RILEY'S KAYAK (bKr–12)

Town Hall, Braintree, Essex, England

length	570.0 cm (18 feet, 8 ½ inches)
width	49.1 cm (19 ⅜ inches) at 275.0 cm station
depth sheer	17.2 cm (6 ¾ inches) at 275.0 cm station
paddle length	206.0 cm (6 feet, 9 ⅛ inches)

Surveyed 1987, Stella and John Brand.

This kayak [Figure 5.6] was used by Lt. Cmdr. Quintin Riley on the British Arctic Air Route Expedition, 1930–1931. It was the first Greenland expedition led by Gino Watkins and a remarkable part of British history [see Riley 1989]. At that time kayaks were still essential tools in the Eskimo economy, and locally built kayaks were used at Angmagssalik because [Watkins] believed in living off the land (1988:8).

Two things still affect us today: their work made flying easier and quicker and, because their kayaking exploits became well known, the bad effects of imitating [East] Greenland kayaks became normal, whether canoeists know it or not (1988:9).

I had my first sight of Quintin Riley's kayak when it was stored and well cared for in the loft of his garage at Shalford, Essex. He told me that it had been Nicodemossee's own kayak and it had been re-covered when [Riley] acquired it during the first Greenland expedition. Thus [Riley] is one of the few Englishmen who have been able to take over a native flat-decked kayak without altering it. This enhances its value historically and ethnographically. I assume that Nicodemossee parted with his kayak because of his advancing years (1988:10).

The paddle is 206 cm (6 feet, 9 ⅛ inches) long and 5.5 cm (2 ⅛ inches) wide halfway along the blade.... [Riley]'s paddle has the same kind of sections as the West Greenland paddle [shown in Figure 4c], i.e., with an elliptically shaped loom so that the greatest depth of material resists the paddling thrust.... Henrik Kaput found out that 'bent' paddles were usual in most kayak areas and deduced that they were more efficient than straight blades and looms. So some of my early drawings may show bent paddles that I ignorantly straightened, not realizing that the 'warping' was not only intentional but technically advanced (1988:14).

A manhole width of 35.4 cm (14 inches) limits the number of people who could use a semi-replica. A number of those would be young so they would probably find the kayak too long and heavy.

Even so, it would be an ideal kayak to reproduce, it is easily the most beautiful of all the kayaks used on the Air Route expeditions and has that sense of completeness that is rarely achieved. It is extremely valuable historically and ethnographically, having had two famous owners and been suitable for both of them (1988:17).

WEST GREENLAND KAYAK: OLD DISKO BAY TYPE (bKr–13)

British Museum, Ethnographical Department

length	535.3 cm (17 feet, 6 ¾ inches)
width	47.2 cm (18 ½ inches) at 260.0 cm station
depth sheer	18.3 cm (7 ¼ inches) at 260.0 cm station
paddle length	206.0 cm (6 feet, 9 ⅛ inches)

Surveyed 1963, John Brand.

This was the first kayak [Figure 5.7] I measured—soon afterwards I teamed up with Charles Ranshaw, the circulation manager of *Canoeing*, as part of that magazine's Project Eskimo, and learnt better techniques (1988:18).

When talking about bKr–13 it has been easy to slip into the habit of referring to it as the 'old' [British Museum kayak] to differentiate it from the other Greenland kayaks in the [museum]. John D. Heath pinpointed its area of origin as "Disko Bay or vicinity" in his letter dated 17 August 1963, but some mystery surrounds its likely date of construction. There is no disagreement that it is an old Disko Bay specimen because after 1880 the ends would have been lower (1988:18).

I hope that the drawing is some indication that the 'old' Greenland kayak is a beautiful, well-conceived kayak. Certainly both the man who made it and the man who bought it knew what they were about. The shape lacks the kind of angularity common all over the world in the [twentieth century], and the flow of the sheer line is subtle and sinuous in a way [reminiscent] of the European Baroque or Art Nouveau styles (1988:19).

My two replicas made in the 1960s kept the flat bottom aft but now I would let the keel line run parallel to the chines. After my experience with a replica of [bKr–9], I would try hard not to increase the aft buoyancy. If the flat area was caused by the weight of the kayak pressing down on the bracket, it is correct to return all the ribs in that area to their original shapes, i.e., pinch in the chines so that the shape is faster. Basically, all that should be done is to spring a stronger line from 2,000 mm and fair it in to the stern somewhere

Figure 5.7
WEST GREENLAND KAYAK (bKr–13)

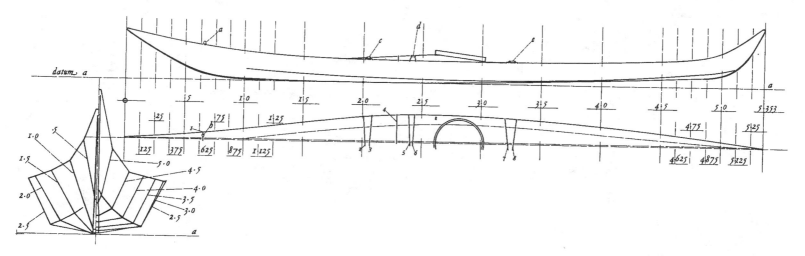

near 4,625 mm. Then raise the chine line uniformly above the new keel line. On the half-plan the chine line should be redrawn from about 2,250 mm so that it enhances the fish-shaping of the chines. I suggest that at 3,500 mm the chine dimension might be tried at 116 mm instead of 126 mm. It is then likely that the chine will end in the area of 4,950 mm instead of 5,100 mm (1988:23).

An actual-size replica would suit most small kayakers, particularly if the construction was monocoque and care was taken to design away obstructions. The higher deck in front of the cockpit gives the Disko Bay type a definite advantage over other Greenland kayaks when considering ease of entry. I was in a semi-scale replica of the [British Museum] kayak when I found that it would obligingly lay on its side after a capsize. A sweep of the paddle would have got a competent paddler upright again (1988:23).

I was very fond of the Disko Bay shape in the 1960s and I still think it represents all that is good in Greenland kayaks. It is neither extreme, in the quirky sense of the word, nor average. The master must have enjoyed realizing what he had seen in his mind's eye: the long curves of the forebody, the strong curves of the stern and its smoothness are a delight and an education.... [A] good semi-replica of the [British Museum's] Disko Bay kayak would show how strong the authentic tradition is and how weak it is to merely imitate some Eskimo shapes (1988:23–24).

EAST GREENLAND: GINO WATKINS' WHITE KAYAK (bKr–18)

Lancing College, West Sussex

length	516.5 cm (16 feet, 11⅞ inches)
width	49.5 cm (19 ½ inches) at 244.0 cm station
depth sheer	19.7 cm (7 ¾ inches) at 244.0 cm station

Surveyed 1987, John Brand.

This is probably the most famous kayak [Figure 5.8] in the U.K. amongst canoeists because of the much reproduced photograph [from *Those Greenland Days*] of Gino Watkins sitting in it [Figure 5.9]. The pose is meant to recall, or imply, many of the qualities in vogue before World War II: youth, cleanliness, vigour, and modesty with a touch of novelty thrown in. There is a direct parallel with chivalry, [Watkins'] indisputable charm being wonderfully reinforced by his white kayak (1988:66).

...I gather that the white kayak was probably the first to be made on the 1930–1931 Greenland expedition. I expect that it was made by [Nicodemossee] (1988:66).

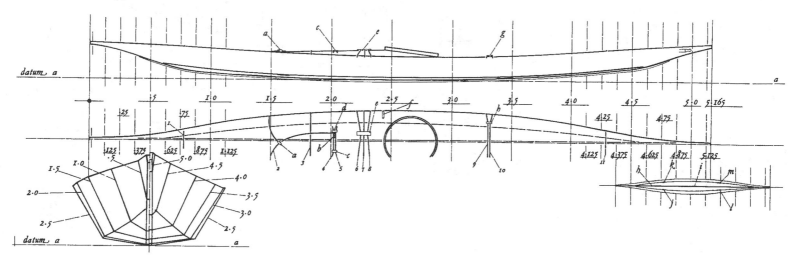

Figure 5.8
WATKINS' WHITE KAYAK (bKr–18)

Figure 5.9
Gino Watkins seated in the White Kayak.
Photographer I. Cozens (?), no. P48/16/21.
Courtesy Scott Polar Research Institute,
University of Cambridge.

EAST GREENLAND: F. S. CHAPMAN'S THIRD KAYAK (bKr–19)

Atlantic College, St. Donat's Castle, Llantwit Major, South Glamorgan, Wales

length	5286 mm (17 feet, 4 ⅛ inches)
width	481 mm (19 inches) at 258.8 cm station
depth sheer	187 mm (7 ⅜ inches) at 258.8 cm station

Made in 1932 for the Second British Arctic Air Route Expedition. Surveyed 1987, John Brand.

The kayak [Figure 5.10] is the only one surviving of the three once owned by F. Spencer Chapman. His main claim to fame, as far as canoeists are concerned, is *Watkins' Last Expedition* (1932). That book, together with some passages in *Gino Watkins* (1937) by F. M. Scott, made [Watkins] a cult figure (1988:71).

In *Those Greenland Days* (1932), Martin Lindsay records on page 180 that the First Air Route Expedition had ten kayaks made, most of them by [Nicodemossee] who charged about £5 for each kayak. Even today this seems to be an enormous sum since it was only after World War II that an experienced brick-layer could expect £5 a week in wages.... Reread today, the passage [from *Those Greenland Days*] still has a directness that gives a good idea of the East Greenland standards then current. Unfortunately there was the failure to mention other types of kayak and the assertion that the kayaks used by the expeditions were the ordinary Eskimo pattern spread.... The canoeing books written by such people described the East Greenland type as the only Eskimo kayak because it was the only one they had in mind or knew of, since they evidently did not visit museums. This is something that will continue to hinder kayak research until a wider view is taken (1988:72).

...page 142 of [*Watkins' Last Expedition*] gives some information about [Chapman's third] kayak:

(a) [Chapman] felt safer in it;

(b) [Chapman] said that stability was regulated by the angle of the sides, not the depth or the width;

(c) it was 50 mm narrower than [Chapman's second kayak];

(d) it had very little free board (1988:74).

A Semi-Replica? Recreational benefits apart, there are several good reasons for making a semi-replica:

(a) Chapman's enthusiasm for the performance of his kayak could be assessed and compared with the performance of the other Air Route kayaks;

(b) the [very] low volume is a challenge as well as being unique: even with monocoque construction some idea of the suffering borne by F. S. Chapman could be estimated;

(c) semi-replicas are a way of checking the validity of restorations made only on paper...;

Figure 5.10
CHAPMAN'S THIRD KAYAK (bKr–19)

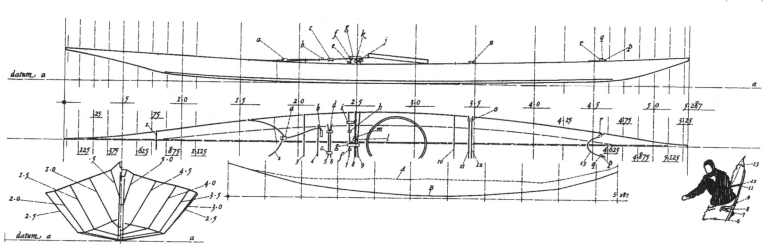

(d) the pleasure of paddling something close to Chapman's original goes without saying and it does seem to be the most advanced kayak specially made for a member of the Air Route expeditions but what [Chapman] wrote about kayaking gives him an outstanding place in pre-World War II history (1988:81).

Only small kayakers would suit a semi-replica if it is to be used safely. Chapman may have been a little larger than some Air Route kayakers if the large hoop and scooped out deck beams are anything to go by.... A first-size hoop would enhance the looks of the kayak and monocoque construction would help entry and exiting (1988:81).

GREENLAND KAYAK (bKr–20)

Danish National Museum, Ethnographical Department, Copenhagen (no. 3)

length	558.8 cm (18 feet, 4 inches)
width	52.4 cm (20⅝ inches) at 2713
depth sheer	17.8 cm (7 inches) at 2713

Surveyed 1975, Stella and John Brand.

I am not sure which part of Greenland the kayak [Figure 5.11] came from, or when it was made. At first I assumed it was East Coast because of its position in the gallery. Later I thought that if it was from the Angmagssalik area it would be late nineteenth century because of the elegant ends. More recently I have thought of South Greenland as its place of origin—not only does it have a fine rising stern but the aft cross sections are fairly boxy. However, its last deck-line is adjustable like its neighbour in the gallery (bKr–21), which is listed as being from Angmagssalik (1988:82).

A Semi-Replica? For the small- to medium-sized expert this kayak promises to be unusually rewarding. The overall shape is elegant compared with most of the Air Route expedition kayaks because the cockpit is small and the ends particularly fine. It ought to be easy to roll. Altogether quite a good-looking kayak even if fairly long and broad for a hunting kayak (1988:86).

Small kayakers may consider reducing the beam to produce a kayak more suitable to their weight; the sheer plan is much more satisfying than the half-plan (1988:86).

Figure 5.11
GREENLAND KAYAK (bKr–20)

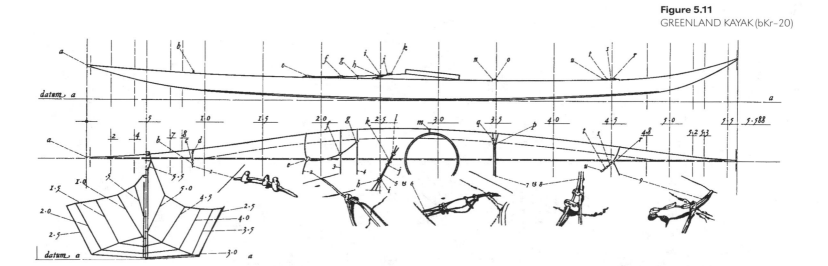

Figure 5.12
EAST GREENLAND KAYAK (bKr–21)

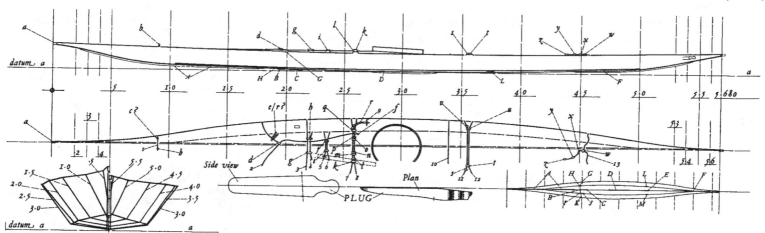

EAST GREENLAND KAYAK (bKr–21)

Danish National Museum, Ethnographical Department, Copenhagen

length	568.0 cm (18 feet, 7 ⅝ inches)
width	47.6 cm (18 ¾ inches) at 2730
depth sheer	15.4 cm (7 inches) at 2730

Surveyed 1975, Stella and John Brand.

This was the first kayak we measured [Figure 5.12] at the [Danish National Museum], it was lying in the gangway of the basement gallery when we arrived and the reason for this is probably explained by Erling Ekegrin's note for it: "Angmagssalik, no number, Knud Rasmussen's kayak, recent acquisition, via Öregaards Gymnasium where the kayak was received as a gift from Knud Rasmussen about 1930, no information as to origin."

It is of interest to note the similarity of some features between this kayak and those built for the Air Route expeditions. It looks as if Knud Rasmussen (1879–1933) was a better judge of a kayak than the Englishmen and was more experienced in getting the kayak he wanted (1988:87).

This is an intriguing kayak; at one level the flat deck, heavy coaming, and a surfeit of keel and chine protection pieces tempt a comparison with high fashion where deliberate gaucheries emphasize opposite trends, e.g., slim, flat ends and very deep protection pieces (1988:91).

With only 150 mm (5 ⅞ inches) clear space under the cockpit main beam, only small kayakers can expect to use a semi-replica. Raising the foredeck would not only spoil the lines, it would make any semi-replica a hybrid. Therefore, it is better, culturally and technically, to choose another type of kayak if bKr–21 is not large enough. For the right weight kayaker, an [authentically shaped] replica would probably prove to be a more exciting performer than many East Greenland craft (1988:91).

REFERENCES

Brand, John

1984   *The Little Kayak Book*, part I. Colchester, Essex: John Brand.

1987   *The Little Kayak Book*, part II. Colchester, Essex: John Brand.

1988   *The Little Kayak Book*, part III. Colchester, Essex: John Brand.

Chapman, F. Spencer

1932   *Watkins' Last Expedition*. London: Chatto and Windus.

Lindsay, Martin

1932   *Those Greenland Days*. Edinburgh: W. Blackwood & Sons.

Riley, Jonathon P.

1989   *From Pole to Pole: The Life of Quintin Riley, 1905–1980*. Bluntisham, U.K.: Bluntisham Books.

Scott, F.M.

1937   *Gino Watkins*. London: Hodder and Stoughton.

6 Kayak Sports and Exercises

H.C. PETERSEN
Edited by E. Arima

This publication in English of a Kalaallit, or Greenland Inuit, view of kayak training, games, and other practices is possible thanks to the generosity of Greenland kayak and umiak *authority H. C. Petersen and Keld Hansen of the Viking Ship Museum in Roskilde, Denmark. The article originally appeared in Danish as "Kajaklege og øvelse" in the museum publication* Kajakker, *edited by Keld Hansen and Birthe L. Clausen (1991:19–28), along with ten illustrations not reproduced here. —Ed.*

While I was studying and registering kayaks, I frequently became aware of their sports potential. I had heard several adult men mention games with kayaks, and in my childhood I had seen many kayak games performed either by individuals or groups of kayakers putting on a show outside the village. I find the games just as fascinating as the technical construction of the kayak. Below I discuss some of the sports aspects of the kayak, hoping that they will fascinate other kayak enthusiasts as well.

HOW IT STARTS As soon as a baby boy is big enough to sit on his mother's lap, the mother makes him play "kayak games." She takes hold of his small hands and makes him "paddle a kayak." She saw her mother do the same with her little brothers, and now she sits with her own son and honors the generations-old play, chanting in a staccato style this verse, which has no melody:

Qajaralak	Little kayak
piorami	paddling toward seal
naakkialeqaaq	throws his harpoon
tik!	tik!
Avatarsi,	Avatarsi,
avatarsi!	avatarsi!

This is the way the verse was recited in Maniitsoq/Sukkertoppen when I was young, but I have heard several versions along the coast. At "*naakialeqaaq*," the mother pulls the boy's arm into a throwing position and, saying "*tik!*," makes him cast his invisible harpoon through the air. His small hands are then held to the sides of his head, and he is made to rock sideways in rhythm with the syllables of the following lines, simulating a bladder float dancing on the surface of the sea while being pulled by a seal. The boy would gurgle with joy during the play. That is the beginning of his kayak training.

BALANCING BOARD Three- to five-year-old boys who are still too young to have their first training kayak play with small harpoons, which in the beginning are nothing but small wooden sticks, but later carry proper foreshafts. Sooner or later, a boy gets a balancing board, which is a board long enough to sit on with the legs stretched out. Under each end of the board is a crosswise leg rounded at the bottom to make the board rock or overturn easily.

The balancing board often becomes a place for the boy to take his meals on. He will be given, for example, a rib, which he holds with both hands and bites into while keeping his balance. Thus he will have learned how to balance by the time he gets his first kayak.

THE FIRST KAYAK

When a boy sits down in a kayak the first time, he is shown how to hold the paddle and how to use it. The first exercises are about balance and paddling. The coach, usually the boy's father or grandfather, observes the placement of the boy's hands on the paddle, which is not to be dipped too deep in the water, every stroke of the paddle, and every movement, correcting all errors immediately.

While the boy sits in the kayak, the coach holds the nose of the kayak with one hand and pulls it out to the water. Standing at the water's edge, he pushes the kayak out as far as his arm will reach, and telling the boy to "paddle to the shore cautiously," he carefully observes how the boy uses the paddle. The boy is repeatedly pushed out until he learns to balance and use the paddle correctly. Only then does the coach dare loosen his grip on the kayak and push it out a couple of meters so that the boy can paddle in by himself.

Early in the exercises the boy has to learn to manage a situation where the kayak heels to one side. He has to hold one of the paddle blades with both hands, leaving the other half of the paddle free. The free blade is placed in the water and moved back and forth with a circular motion in the water's surface. This way the kayaker is able to tilt his body sideways so much that he almost touches the water yet does not capsize, and then straightens up again. Naturally, the exercise is performed to both left and right.

GETTING IN AND OUT OF THE KAYAK

The Greenlandic kayaks are very narrow and shallow. To get into and out of his craft, the kayaker has to bend his legs backward. This is why a kayak hunter, on land, is characterized by knees that bend backward and lower leg muscles that stand out. When getting into a kayak, he seats himself on the backmost part of the cockpit coaming and then lowers himself into place by squirming.

Sometimes getting in and out of a kayak fast is a matter of life or death. The kayaker may be surprised by a sudden storm and have to get ashore fast, or he may have to go to someone's rescue. Or he may have to make shore between two breakers, quickly bringing the kayak and himself to safety before the next breaker.

These skills are put to the test in an enjoyable form of competition: at a signal, kayaks lying on shore about 10 meters (33 feet) from the water's edge are grabbed, brought to the water, and paddled to an adjacent peninsula or small island, where the competitors land, portage their vessels to the other side, and then paddle back to the start.

KAYAK EXERCISES AND COMPETITIONS

Racing is one of the most common competitive kayak exercises. In another game, kayakers compete to see who can hold his head underwater the longest. Boys start to practice for this early, holding their breath in a bucket of water to test how long they can hold their breath. Later, when they have learned to do the kayak roll, the competition is about who can stay underwater longest with his head pointing down.

When I was a boy, I heard a tale about some friends who had grown up together in Kangaamiut at the end of the last [i.e., 19th] century. They had learned together to paddle, and when they went hunting in Davis Strait they took every opportunity to practice kayak games and compete. Once, when they saw some king eiders (*Somateria spectabilis*) dive for mussels, they turned their kayaks upside down. One after the other they had to right themselves, until at last only one king eider and Kalistat stayed immersed. After the bird came up, Kalistat slowly righted himself. I have heard some of these friends claim that Kalistat was the champion of this sport.

"Up and over a kayak" is a game where one kayaker stays put, usually with his hull turned up, while another attempts to paddle over the upturned kayak. If the speed is high enough, the kayak almost "jumps" over the other craft, while insufficient speed results in a risk of stopping and overturning.

Surf-riding is performed fully clothed in fur and with all equipment removed from the deck. The kayaker headed either directly against the breaker or at a slant — a balancing exercise mastered by only very few kayakers.

The Great Rock Off Akuliaruseq. The now-deserted village of Akuliaruseq lies a few kilometers south of Alluitsup Paa, in the municipality of Nanortalik. An elongated rock ending in a steep wall is reported to be found in the vicinity. When storms from the southwest rage for several days, the great Atlantic waves grow huge on their thousand-kilometer path, and the swells carry an enormous force. During a storm, these waves pound Greenland violently. It is not wise to set out in a kayak. Only the strongest and the most skillful kayakers could handle their craft in the huge Atlantic waves; however, the kayakers of this village were reputed for their daring, and those who felt like braving the violent storms wore full fur suits made of unfleshed harp seal (*Phoca groenlandica*) skins. (The fleshed skins were reported to be inclined to rip.)

Storms were followed by huge swells that formed breakers over the rock. The kayaker positioned himself in front of the rock to wait for swells, which formed deep troughs in the sea before him. He paddled up to the crest of a swell and rode the crest. The wave carried him forward until the powerful surf stopped him at the rock and he was dropped down after passing the back wall of the rock.

The kayakers competed with each other to fall with the wave with their head pointing down, readying the paddle to be able to arrive in an upright position at the back of the rock. Only the most skillful of them, however, were able to end up in an upright position, and most of them are reported to have had to right themselves afterwards.

Exercises with Bladder Float and Line. In another game, the kayaker attempts to push his bladder float underneath his kayak. This is usually done close to the shore just outside the village. The kayakers wear full fur clothing, and everything except the paddle is removed from the deck. The paddle is placed under the cross straps so that it is ready for use.

It is very difficult to push a fully inflated bladder float into the water. Therefore some of the air is let out. The kayakers compete in who can keep the most air in his float. When the float is pushed under the kayak, the kayak is lifted up, its center of gravity is disturbed, and balancing becomes very difficult. Most kayakers can expect a dip in the water, but if the paddle is kept ready, those who know the roll stroke are in no danger.

"Being pulled sideways" is performed with a long line, one end of which is held by boys and young men standing on the shore. The kayaker paddles out. All equipment has been removed from the deck, and the kayaker wears a full fur suit. The line is then taken either over or under the kayak, and upon a signal, the men begin to haul in the line or to run inland away from the shore.

While being pulled sideways, the kayaker has to keep his balance and reach the shore without overturning. This game is also played by taking the line underneath the kayak and then tying it to a fully inflated bladder float. This is a situation that may occur if the kayaker harpoons a seal and does not release the float. Daring hunters used to train themselves in not throwing out the float and hauling the harpooned seal in. This was a dangerous game not dared by many, as the seal could be so strong that the hunter would perish just by the tension of the line.

"To slip under the line" can be played where current meets waves and big waves are formed by fresh wind. This is played in a fjord with a strong current or out at the banks in Davis Strait. I have heard this game mentioned in several places. A line, if stretched between a harpooned seal and a bladder float, tightens to form a bridge between the crests of two big waves. This is a very tempting situation for the other hunters, who attempt to scurry through under the line between the wave crests. It can be done, but those who are hit by the float can count on a spin in the water.

OTHER KAYAK GAMES A number of games are played at home or on the cliffs, intended to be useful for a hunting kayaker by helping him to train the muscles in his fingers, arms, hands, and his entire body:

> *Qiterlimminneq*—to pull fingers;
> *Pakassumminneq*—to pull arms;
> *Arsaaraq*—to pull on two handles tied together with a short strap;
> *Eqitaq*—to pull on a smooth round stick 25 to 30 cm long.

Line Sports. A number of games are played with either one or two stretched lines. They are intended to train a kayaker to right himself upon turning over in his craft, which requires great strength and agility.

Qajaasaarneq, the kayak game, is played with two lines stretched parallel. The player gets up on the lines and places one leg on each line with his behind between the lines. The left hand is taken under the left line and grabs the line on the right, and the right hand is placed correspondingly. When the person turns over to the left, he pulls simultaneously on the right line with his left hand and this movement makes him roll up on the right side, and vice versa.

Qipineq, rolling, is a continuation of *qajaasaarneq* and is played on the same lines; however, in this game the turn is performed several times in the same direction until it is no longer possible to roll. While rolling, the person's hands must not let go of the lines.

Nappaaneq is performed with a single line tightly stretched at shoulder height. Grabbing the line with both hands, the person lifts himself up so that his stomach rests on the rope and somersaults over it.

Kimmimminneq is played so that the person hangs from the line. Holding it with both hands and catching it with his right heel, he attempts to get on the line and roll around it.

Oqaatsuaasaarneq refers to hanging from the line in such a position that the person holds one hand to his head and the other one under his belly. He must then catch the line with one foot and roll around in this position.

And finally there is *tunussineq*, in which the person hangs from the line with his back pointing down, grasping the line with one hand under his neck and the other under the small of his back. It is possible to roll around in this position as well.

COURTING

A young girl sitting at an oar in an *umiak* had her eye on one of the young kayakers keeping close to the *umiak*. During the trip the hunters were casting their harpoons for fun. The young man, raising his harpoon suddenly, cast it toward the ring formed by the girl's oar on the water surface: the harpoon landed in the middle of the ring with a splash. The young girl blushed, and the others shrieked with delight. What they had just seen could not be misunderstood: the young man had confessed his love for the girl.

KAYAK SPEED

Kayakers able to paddle at high speeds have always been remembered and mentioned with respect. The speed of a kayak is naturally determined by several factors: the paddler's strength, his technique, and the kayak's construction and shape. Many reports mention kayak tours, long-distance paddling, and paddling speeds, but it is not possible to calculate speeds on the basis of these reports. One day I heard a kayaker on Radio Greenland tell about his kayak trip from Igaliko to Frederiksdal, a distance of 185 kilometers [115.6 miles]. He had departed early in the morning and reached his destination in the evening. My grandfather once paddled on open waters from Kangaamiut to Sisimiut, a stretch of 136 kilometers [85 miles]. He started at 6 AM, and the colony clock struck 6 at his arrival in the evening. Thus his average speed on this trip was 11.3 kilometers per hour [7.1 mph].

Kussarsarluni paarneq, that is, paddling with a lowered stem, is the technique most commonly used to make good headway. Ezekias Davidsen from Tasiusaq, Nanortalik District, who was known for high-speed paddling, described this technique as follows: "The paddle is dipped deep into the water, and the stroke is performed fast with an upward motion so that the current hits the aft end of the kayak from below and lifts it upward. This makes the bow tilt down a bit. The angle of the paddle blade is changed during the stroke by a wrist movement. The strokes are swift with short intervals."

The distance between Tasiusaq and Nanortalik is 27 to 28 kilometers [16.9 to 17.5 miles]. Ezekias Davidsen could cover this distance in an hour, sometimes in 55 minutes. He explained these exceptionally speedy trips as follows: "It happens when I run out of tobacco and have to get to the store fast." It must be mentioned here that an ordinary cutter spends two and a half hours on the same trip. Thus *Kussarsarluni paarneq* may produce amazing results.

INSHORE USE OF KAYAKS

Transporting a kayak overland on top of one's head is known as *maqinneq*. Over shorter distances the kayak can be transported by carrying it with one arm inserted into the cockpit to grab one of the deck beams. If a kayaker carries his kayak on top of his head, the bow is made to point backward because it is often the heaviest part. The hunter rests the kayak on his forehead, and if he wears a piece of soft skin as protection against sharp edges, he does not have to use his hands at all.

The hunters used to take their kayaks with them on caribou hunting trips in the fall, when they moved with their families into the fjords in order to get to the hinterlands where they frequently had to cross lakes and rivers. The kayaks were used as ferries, two of them frequently being lashed together with some pieces of timber. The passengers sat on the fore- and afterdecks.

The kayaks were frequently carried all the way to the inland ice. The ends of the fjords often became known by the name *umiivik*, indicating a place where the women's boats were left behind, while the name *qajartoriaq*, Kayak Lake, became quite common inland. The places where kayaks were portaged from fjord to fjord were called *itinneq*, *itinnnera*, etc., meaning Portage Place. By using the portage places, long detours of many kilometers were often avoided.

7. A Dramatic Kayak Trip, 1899–1900: Ataralaa's Narrative

JOHANNES ROSING (ATARALAA)
Edited by E. Arima

The following account of survival aboard a kayak during a storm appeared in Danish as "En dramatisk kajaktur ved årsskiftet 1899–1900" in Kajjakker, edited by Keld Hansen and Birthe L. Clausen (1991). The account was recorded at the request of Knud Rasmussen. —Ed.

One of my very first childhood memories is of my small kayak. I do not remember exactly when I got it, but I remember from the time on when my memories began to become coherent how much I liked to use it. I was given kayak tools before my confirmation, and from that time on I started to train myself in the things I saw hunters do. After I finished school, I began serious efforts to learn to paddle out at sea. I stayed at home only during the very worst weather.

When I was 22, my father got me a job as the manager of a trading station. In spite of the permanent employment I became increasingly eager to perfect my skills. I taught myself coopering. Also, whenever possible I went hunting by kayak with other hunters, always returning home for the night. My wages as a cooper were small, and hunting was at times more profitable, which was very important to me.

During my 13 years as manager of the trading station, I had gradually developed a growing debt that I had to pay back to the Royal Greenland Trading Company. Therefore, I requested the company to let me resign and become an ordinary employee. My request was granted, and I moved to Maniitsoq [Sukkertoppen], where I was employed as a cooper. I did not give up hunting by kayak and spent my spare time hunting to provide for my family.

The years went by. The 31st of December, 1899, was a Friday. It was a day that made me very happy. My family traditionally invited each other for coffee in the early morning. I lay in bed and waited for the morning and the invitations.

The weather was clear with a weak westerly wind and seemed stable. I got up before dawn and walked down to my kayak, which was at some distance from the village because a few days before new ice had forced me to shore west of the village. I paddled the kayak out to look for seals in this area where I used to hunt with my harpoon.

I paddled out alone but later encountered another kayaker, who did not, however, wish to go farther out. I went on alone, and the seals were there, but they were very restless and therefore it was difficult to get within harpooning range. The waves splashed over the kayak noisily. In the afternoon I turned back and started to paddle home. I wanted to return in time for the evening service.

On my way home I saw that it was snowing heavily behind me in the west, but I did not hurry because the weather looked so fine. Close to land the snow caught up with me, while the wind continued to abate. There was almost no wind by the time my vision became obstructed by the falling snow. Some time later I had to paddle around an area with new ice, and as I was paddling I suddenly sighted a hooded seal in front of me. I wanted very much to get it but hesitated. Then, however, I decided to harpoon it and started after it at a great speed. I wanted to hit it in the back, in the spine, so that it would die immediately, but failed. The seal dove, and I threw out the float and picked up the harpoon shaft. The float was pulled under the water. I investigated the thickness of the new ice and found that it could not be forced. I had to paddle back, and even if I were to see the seal again I would have had to give it up because it would have been impossible to tow it to shore through the new ice. I had to try to reach land before it got too dark.

By this time I had become disoriented, although I was not aware of this immediately. I paddled and paddled, and it became dark. I knew that the congregation expected me to turn up at the evening service. I tried to hurry on but to no avail. The wind was rising, and it got increasingly warm. I stopped paddling and listened but could hear only the sound of the waves.

After paddling a long time without sighting land, I turned back, believing that I had passed Maniitsoq. I continued paddling and began to fear that I would perish. I carefully tied the openings of my kayak fur jacket and secured my weapons so that they would not fall out. I was frightened when I realized that I was facing the dangers of the dark night in the storm. Even if I did not perish in the storm, it would be an unpleasant experience. I was being continuously hit by the breaking waves and was afraid of losing my paddle because then I would be lost. I could hit a rock or be thrown against the shore, which would break open a hole in the kayak's skin cover and then I would turn over. The thing I was most afraid of, however, was that the sea would get me alive. I shuddered to think that my wife and three children would be left alone and homeless. At that time we were living with another family and I had not finished paying my debt. This was what made me really sad. I decided not to surrender to death; I would use all my strength to save my life as long as the strength was still in my possession. I turned, after the Christian tradition, to He who can help in all situations and asked Him to give me sufficient strength to pull through.

I exerted myself until I was thoroughly exhausted and became sleepy. This was the worst part of it. As I tried to keep my eyes open in the darkness, I began to dream I was in a warm house playing happily with my children. Some time later I discovered that all was cold around me and I realized that I had overturned. I succeeded in righting the kayak and was surprised at not having lost my paddle.

The whole night passed in this manner, and I had to use the paddle continuously to keep my balance. How fortunate I was to know the kayak roll so well! Never again was I to live a night as long as that one. The luminescence of the sea made me feel insecure.

While I was struggling, a couple of long-tailed ducks flew past. I realized I wasn't the only living being out there on the windswept, heaving sea. Their shrieks sounded wonderful, as if they were the voices of many people. While I was struggling to hold myself upright

and to turn around, drifting perpendicularly to the wind direction, suddenly I heard a whistling sound. It sounded like a stick being thrust into dry snow. The sound came nearer and I heard it behind my back. The kayak capsized again, and when I righted, the sound grew more distant. I emptied my mouth and eyes of water, and shortly after I was able to feel the wind in my face and turned toward the wind. I was very cold, and the snow crystals made my eyes hurt. I could not stay with my back to the wind, however, and therefore placed myself athwart the direction of the wind.

When I turned my back to the wind, I wondered whether the wind was actually changing because now it blew into my face. I turned again and wondered whether I had come to the lee of the land. I could distinguish something white in front of me. It very much resembled the white church of Maniitsoq, and I thought that I must have come to the leeward side of the church. However, it did not have any windows, and the walls bulged out where the windows should have been. I collected my thoughts a bit and tried to touch it with the bow of my kayak. It was nothing. But still I could distinguish the foremost part of the kayak in the light reflecting from the church. This I found most remarkable, and as I paddled farther out I could see the roof, which was also white. I paddled back and attempted to touch it but again without result. I drifted to one side of it, reached the corner, paddled back, and drifted farther. I completely forgot my worries and stopped thinking of my family; no longer was I afraid of dying.

While I was staying there I realized that the sea had become very turbulent. I thought I saw a weak light in the west, but it turned out to be the east, and no land was in sight. The big waves could not possibly come from the land. I paddled fast to get warm. I was still in an area with a raging southwesterly storm but I was so relieved that I said to myself, "There is coffee there waiting for me."

While paddling I got out of the snowstorm, and now I could see land. I thought it was Naajarsuit, an island west of Maniitsoq, and directed myself toward it while paddling as fast as I could. I was wringing wet but it made no difference how fast I was going; the distance did not seem to get shorter. As the day lit up the distance was still the same.

Then the blue sky dove down above the land, and I recognized a high mountain at the head of Kangerluussuatsiaq Fjord [Evighedsfjord]. I know that mountain very well, since I am originally from Kangaamiut. Now I understood that I was far out in Davis Strait, west of Kangaamiut, which is 54 km [33.5 miles] from

Maniitsoq. This was how far I had drifted during the night. Now I could see the outermost islands and encountered the easterly wind of the fjord, while the wind out at the coast was still from the southwest. That was a dire disappointment because now I had to paddle against the wind. However, I had to go on regardless of how exhausted I was. The great rock at Qasissat had to be passed by the windward side, which I succeeded in doing. Close to the outermost island I got a strange feeling: I felt that my senses and thoughts were failing me, but fortunately this lasted only a short time.

I reached the shore finally and arrived in Kangaamiut just before dark. At that time my parents were still living there. It made them both scared and happy to see me. I was saved. My bird-skin under-jacket was completely ruined, and it was great to put on dry clothes. My parents did not know what to do, but I said, "I am out of shape now, Dad, and famished. Give me something to eat." And thus I spent the night of the 1st of January, 1900, with my parents. It felt wonderful, after an absence of 15 years, to celebrate the first night of the new year with my aged parents. When I went to bed that night I could not fall asleep. Continuously I saw in front of my eyes my weeping wife and children.

Finally the morning was near. I did not hear the chiming of the clock, got up at 5 AM and departed around 8. We paddled out before the day dawned, my younger brother, another hunter from Kangaamiut, and I. The weather was fine even though the wind had not fully died down. We reached Maniitsoq at nightfall. My exhaustion had delayed us.

There was great joy at our arrival in Maniitsoq. People crowded around me when I stepped ashore. They shook my hand or fired welcoming shots. At home I found the entire house full of people. I had to push my way to my wife, who sat on a stool paralyzed, not moving at all. She was upset and still in doubt. I broke out in tears as I saw our children, and I said to my wife, "God has given me to you a second time." It was a great moment, to meet again the person for whose sake I had struggled against the storm.

The people of my village now began to celebrate the New Year. The day before everything had been ruined because I was missing. They had waited for me and fired a gun to help me find my way. I had not heard it because of the storm. I had not noticed the turning of the wind and had paddled farther and farther.

I spent the next day in bed. My pelvic region ached badly and two lumps developed on each side. For several days I walked with great difficulty because of the pelvic pains.

That is my report of a kayak trip.

PART II

The East Canadian Arctic

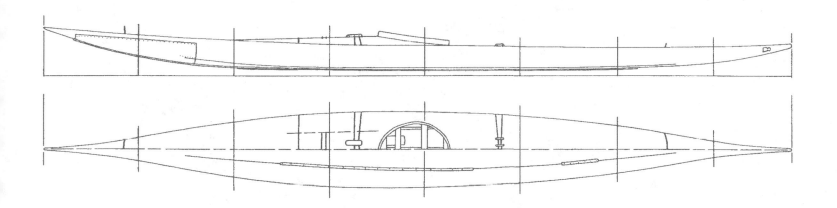

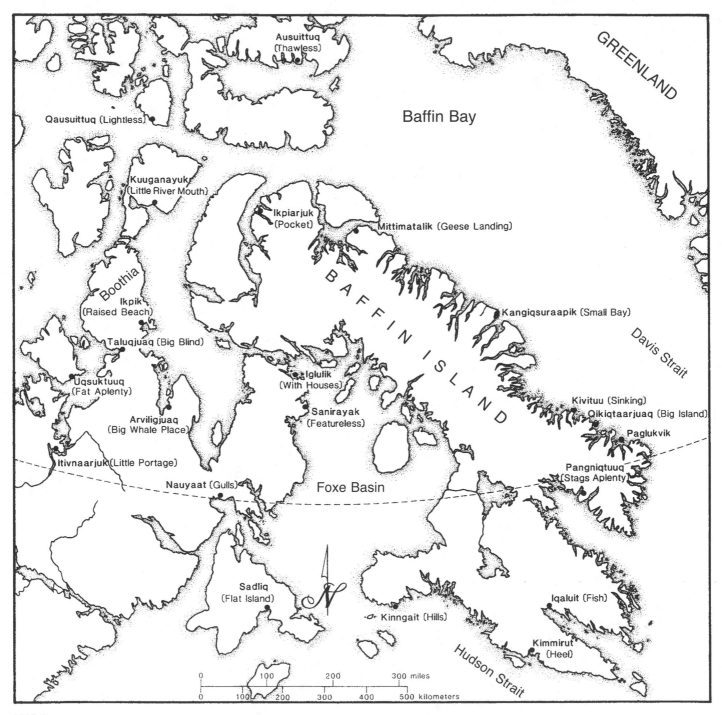

GREENLAND

Baffin Bay

Ausuittuq
(Thawless)

Qausuittuq (Lightless)

Kuuganayuk
(Little River Mouth)

Ikpiarjuk
(Pocket)

Mittimatalik (Geese Landing)

Boothia

Ikpik
(Raised Beach)

Kangiqsuraapik (Small Bay)

Davis Strait

Taluqjuaq (Big Blind)

BAFFIN ISLAND

Uqsuktuuq
(Fat Aplenty)

Iglulik
(With Houses)

Kivituu (Sinking)

Qikiqtaarjuaq (Big Island)

Arviligjuaq
(Big Whale Place)

Sanirayak
(Featureless)

Paglukvik

Itivnaarjuk (Little Portage)

Pangniqtuuq
(Stags Aplenty)

Nauyaat (Gulls)

Foxe Basin

Sadliq
(Flat Island)

N

Iqaluit (Fish)

Kinngait (Hills)

Kimmirut
(Heel)

Hudson Strait

0   100   200   300 miles

0   100   200   300   400   500 kilometers

MAP 5
Baffin Island and the Foxe Basin.

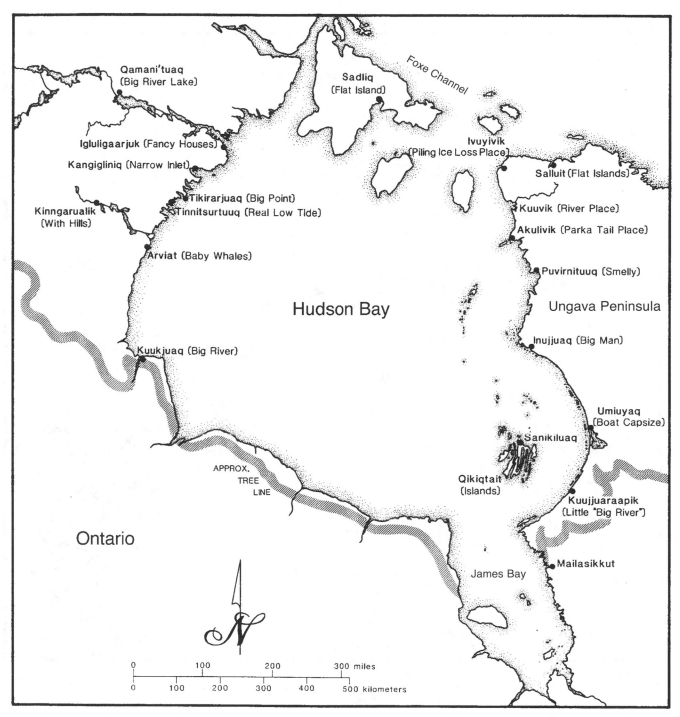

Qamani'tuaq
(Big River Lake)

Sadliq
(Flat Island)

Foxe Channel

Igluligaarjuk (Fancy Houses)

Ivuyivik
(Piling Ice Loss Place)

Kangigliniq (Narrow Inlet)

Salluit (Flat Islands)

Tikirarjuaq (Big Point)

Kuuvik (River Place)

Tinnitsurtuuq (Real Low Tide)

Kinngarualik
(With Hills)

Akulivik (Parka Tail Place)

Arviat (Baby Whales)

Puvirnituuq (Smelly)

Hudson Bay

Ungava Peninsula

Kuukjuaq (Big River)

Inujjuaq (Big Man)

Umiuyaq
(Boat Capsize)

Sanikiluaq

APPROX.
TREE
LINE

Qikiqtait
(Islands)

Kuujjuaraapik
(Little "Big River")

Ontario

Mailasikkut

James Bay

0   100   200   300 miles

0   100   200   300   400   500 kilometers

MAP 6
Hudson Bay.

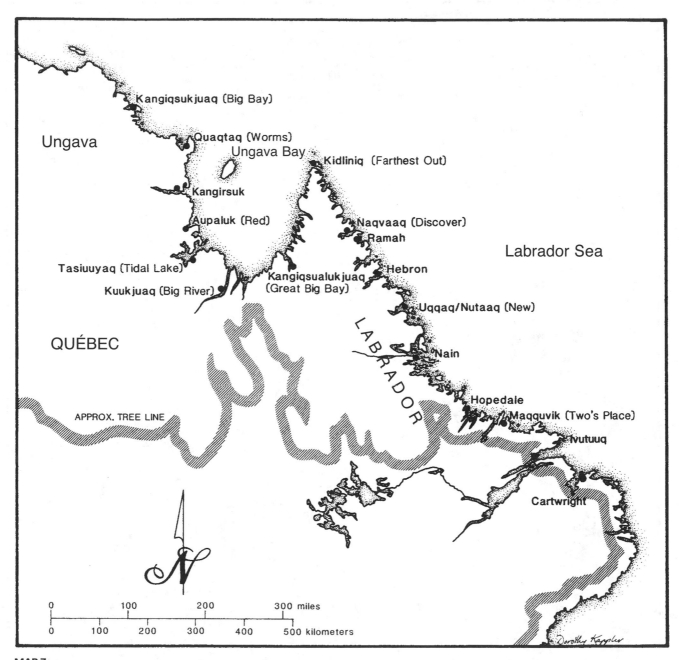

MAP 7
Ungava Bay and the Atlantic coast of Labrador.

8 Kayaks of the East Canadian Arctic

E. ARIMA

INTRODUCTION — A flat bottom, a great well-raked cutwater prow with a deep forefoot or bow bottom, and a low, shallow stern form the basic pattern of the East Canadian Arctic kayaks. There is a strong flare to the sides, which may bulge a bit but may be taken to be essentially slablike. Maximum width is usually at or just behind the kayaker, who already sits aft, the cockpit front being located amidships or a bit farther back. Compared to most other arctic designs, the cockpit opening is longer so that the kayaker is located well aft in terms of the overall length. But when the overhang of the big bow is taken into account (Figure 8.1), the location is not that extreme. The cockpit front is moderately raised, and the coaming, or cockpit hoop, is tilted back. Earlier, before the late 1800s, the rise and tilt were only slight, as in Greenland kayaks. (Entry and exit then had to be with straight legs rather than bent knees.) In the hull plan outline, the bow stays quite narrow initially. Then the sides have a little concave curve as they begin to widen, becoming convex around the fullness of the hull to the front of the cockpit. The hull continues to fill out around the cockpit and stays full aft, though shallow in profile. Howard Chapelle has explained the performance benefits of "the combination of deep forefoot and the greatest beam well abaft the midlength" of the East Canadian Arctic design as follows:

> When paddled, the craft always trims so that the kayak draws the most water at the fore end of the keelson and the bottom of the stern is usually just awash. This makes the bottom sweep up from the forefoot in a very slight gradual curve to the stern, when the boat is afloat. As a result, the kayak may be said to be of the "double-wedge" form that has been popular in fast low-powered motor boats, since having the beam far aft gives to the bow a wedge shape in plan, while the deep forefoot and shallow stern produce an opposite wedge in profile. It would appear that this form has been found by trial and error to produce a fast, easily paddled rough-water kayak in an otherwise heavy hull (Adney and Chapelle 1964:205–206).

Not only does the East Canadian Arctic group have the largest kayaks, with the hull reaching about 27 feet by 27 inches (820 by 69 cm), it also has the heaviest, with weight ranging up to an extreme of nearly 150 pounds (68 kg), though usually considerably less, say, 65 to 120 pounds (30 to 54 kg). Size varies greatly, ranging from 13 to 28 feet (4.8 to 5 m) in length and 20 to 33 inches (51 to 84 cm) in width, though most of these kayaks fall within the range of 16 to 26 feet (5 to 8 m) by 22 to 29 inches (56 to 74 cm). The upper width figure is from a reconstructed Belcher Islands two-seater (Guemple 1966). This double East Canadian Arctic kayak was a very limited variant. It may have been the result of stimulation by a World War II serviceman at Great Whale River who knew about multihatches in the Aleutians and/or southern Alaska. Only four or so existed, and only briefly (Freeman 1964:70–71).

The hull depth is typically about 9 inches (23 cm) at the back of the cockpit, the place where depth to sheer, or the gunwale, is usually taken for Eastern Canadian Arctic kayaks and given in the caption on the lines drawings. At the forefoot, the hull is about an inch deeper than at the cockpit, while the bow rises to about 14 inches (36 cm) on long hulls and 16 inches (41 cm) on shorter ones, around 18 feet (5.5 m) or less. The stern is about half as high. These end heights are reckoned as the kayak sits on the ground. Once it is in the water, the bow is not so high relative to the stern, since the hull rides deeper forward than aft, unless it is loaded down aft.

The paddle is about half the length of the kayak. The blades are narrow, often under 3 inches (8 cm), for less windage, or area catching the air. If a comparatively short kayak is used with too long a paddle, its tendency to swivel will be accentuated. The paddle is used resting on the coaming, sliding along the forward edge, unless silence is needed. Formerly, when the cockpit used to be sealed in with the gutskin jacket bottom, the paddle would not normally have been rested on the coaming. The skin covering was made of the skin of ringed (*Phoca hispida*), harp (*Phoca groenlandica*), or bearded seal (*Erignathus barbatus*), the last being the strongest. Bearded seal was generally the preferred kayak skin when enough could be obtained above the requirements for lines and boot soles. Since great pressure is exerted on the framework by a bearded seal cover as it shrinks on, the construction has to be heavy enough to withstand the force. More recently, canvas covers with coats of oil-based paint for waterproofing have been used.

In the mid-20th century, East Canadian Arctic kayaks began to fall into disuse, giving way to square-stern freight canoes with outboard motors. Since the passing of these kayaks, a handful were built sporadically over the next generation for sale to non-Natives and to institutions and, more rarely, for individual recreation by a younger person who had missed the tradition of kayaking. A score of reconstructions were built in the 1980s at Igloolik (Iglulik), in the northwest Foxe Basin north of Hudson Bay; Mittimatalik (formerly Pond Inlet), north Baffinland (Map 5); and Inujjuaq (formerly Port Harrison), on the east side of Hudson Bay (Map 6), with some external stimulus and funding, notably from Canada Manpower. These reconstructions were covered with cloth and fiberglass, except for a couple at Mittimatalik, which were covered with sealskin. Use has been limited to recreation at Igloolik and Mittimatalik. At Inujjuaq a half dozen were made. A couple serve as floe-edge hunting retrievers at Kangirsuk (formerly Payne Bay) on west Ungava Bay (Map 7). Today, kayaks in the East Canadian Arctic are predominantly plastic imports from the south, mostly paddled by non-Native newcomers.

In 2000, one of the Inujjuaq reconstructions was copied in fiberglass by a boatbuilder in Kuujjuaq (formerly Fort Chimo) and sent to Kangirsuk for trials by kayak revivalist Zebedee Nungak. Originally from East Hudson Bay, Nungak has this short, wide model in his roots but is aware of its limited distribution. He hopes to get Makkivik Corporation, the Native funding body, to agree to mold the more common long type. The classic historical image of the great 22-to-26-foot sea kayak may yet reappear on East Canadian Arctic waters.

KAYAK HUNTING AT SEA East Canadian Arctic kayaking is poorly reported in the literature. We are fortunate to have recently recorded Native information from the Igloolik Oral History Project run by the Inullariit Society in collaboration with the Igloolik Research Centre. These riches exist thanks to the encouragement of John MacDonald, the centre's manager. The original interviews on audio cassettes, together with their transcripts and translations, are held in the society's archives, and copies are curated in Yellowknife in the Northwest Territories Archives, Prince of Wales Heritage Centre. Similar records exist at Mittimatalik from a 1985 reconstruction project (Arima 1991:118). Much of the Igloolik data directly relate to North Baffin, but the information generally holds for the East Canadian Arctic as a whole. Selected excerpts from several informants are given, lightly edited for economy and smoothness. The translations reflect some of the original mode of expression. Titus Uyarasuk (c. 1908–1984) of North Baffin relates how he learned to hunt *ujjuk*, the big bearded seal, and later *qilalugaq*, narwhal (*Monodon monoceros*):

Figure 8.1—Two kayakers in the Foxe Islands, Baffin Island Expedition, 1926. Note the length of the bow overhang. Photographer: Joseph Dewey Soper. © Canadian Museum of Civilization, no. 69083.

Nutaraajjuk used to teach me how to make things and he gave me all the knowledge needed to understand the behaviours of the animals that we hunted. At times we would hunt together, so when heading back home he would sometimes harpoon a seal and kill it with an *anguvigaq* [lance with breakaway ivory foreshaft]. As we headed for home we would go side by side and he would tell me things that I needed to know. All this time there would be no sound at all from the paddle as he kayaked onwards.... I stayed with Nutaraajjuk most of the time; he was my brother-in-law. He told me that when I harpooned *ujjuk* on the side facing the direction of the kayak, it would splash towards the kayak when the seal skips about. He told me that if I struck it away from the direction of the kayak it would not do any harm. When the harpoon line is attached to the seal, if it does not disappear into the water but rather gets only partially submerged that means the *ujjuk* is ferocious. He went on to say that when faced with a situation of this nature one must backpaddle. So one day I struck an *ujjuk*, and at once the *avataq*, sealskin float, started to go into the water. But it stopped submerging when it was only half under so immediately I paddled backwards, and sure enough the *ujjuk* surfaced between the *avataq* and my position.

I studied the behaviours of the sea mammals. I have never been able to strike *qairulik*, harp seal, but my late younger brother Ulaayuk used to succeed in catching them. But me, I never got to learn how to hit *qairulik*....

As for *qilalugaq*, narwhal, you could do anything you want to do with them as they are not capable of twisting their neck and cannot look backwards. When they start to flee they will go on a straight line so when they surface for air you can follow them from behind. We were reminded that as we followed them we should be careful that we do not get on top of the fluke with the *usuuyaq* [bow]; otherwise, they would just dive deeply and get away from you. As you watch the narwhal starting to surface you let go as soon as it surfaces. You have all the target right in front so you can strike anywhere you want.... When I was going to strike with the harpoon I found that when the kayak was going fast it was much easier to throw the harpoon at a more distant target. Of course it was different when the *qayaq* was stationary. I found that when I harpooned a narwhal it was best to harpoon it around the fluke area. If it was harpooned in the main body, the skin was much softer so that the harpoon head would just rip away....

When I started out for the first time I used to hit the paddle with the harpoon when I was going to aim, before I struck. What I had to do was to grab hold of the harpoon and bring it up above the paddle, reaching out to the latter's end so that the harpoon line does not get tangled with the paddle. When I learned the best position to take I used to strike at will. At first I would throw the float to the water as soon as I made a strike, then I discovered that by turning sideways as soon as I struck there was no need to throw the float as the animal that was attached to the line would just pull it to the water....

The float would be on the surface of the water, and when the narwhal has been killed it sinks so the only thing that allows you to get it is the harpoon line with the float. There are also some narwhals that can stay afloat even after they have been killed, but if the narwhal is a fully grown male, he sinks much more readily on account of the tusk. I used to choose the ones with the longest tusks. The darker narwhals had a powerful pull, much more than the others; this was largely due to the fact that they are young narwhals, while the ones with the spots all over them would slow down much faster. It was important for me to know them, so what I found out was that when the narwhals had spots all over them they would tire more easily as they would be old.... When there's a pod, a lot of them just stay afloat.... As I went after them while they floated, knowing that they were not aware of my presence, they would soon start to dive. When that happened it was most probably because there had been some whales that were submerged and saw me from below as I approached the rest of the pod; so the one that was submerged would warn the rest. That is the way it was. But when there were none submerged one could get right among them. I guess they have the capability to communicate with each other even when they are submerged.... When the narwhals were struck and the harpoon line attached, the narwhals could not go straight; they would go in circles. That was how the narwhals behaved to me. As for my younger brother, whenever he struck one they would go in a straight line....

When we were carrying the float we did not have it inflated all the way. But as soon as you saw an animal, then you would have to inflate to your liking. This is so that if there is wind the chances of it getting blown away is minimized. When you have to go through waves, then use a small string to tie down the float. Along the skin of the kayak you will see loops sewn on so that they can be used to anchor anything that needs to be tied down. These are also used to tie down meat. These loops would be sewn to the skin all the way to the stem [a local feature]....

When I caught a seal I would skin it and load it on including all of the entrails. It was relatively easy to carry your catch when the kayak was not too short.... I used to catch a few seals, and when I had to tow more than one seal I would have to tie them to a thong, same as you would with fish. When you start to tow them it's a little hard to get them going; the same also applies to fully grown male narwhal. You have to paddle hard just to get going, but once you have started, the speed is constant. If I were to try that today I would probably drop dead....

There were times when I had to go through the waters when it was rough; of course, this was usually an unplanned experience. The wind would start to blow while I was out on a trip so I would have no choice but to go through with it. There were times when there was so much white cap that the water would spray as if it was drifting snow. Sometimes the waves would get so high that I would have to climb onto a wave and down again. When one went to the top you would have to push with your paddle and when you hit the bottom you get the back swirl of the wind with a wave in front of you. Faced with situations of this nature there were times I could not help but feel weary about the whole experience....

When the water was really calm, I would get sleepy so I would just go into the hold, lie down, and get a good sleep (Uyarasuk 1990, IE-110, translator Tapardjuk).

This passage on kayaking in rough conditions attests to the seaworthiness of Uyarasuk's design, which seems short for the region at about 17 feet (5.2 m) but preserves an earlier form. A North Baffin kayak collected between 1819 and 1820 at Kangiqtugaapik, or Clyde Inlet, by Captain Parry and now in Exeter, England, is of the same design with a bit more sheer (Figure 8.2). Another North Baffin specimen in Edinburgh, Scotland, is also similar; its provenance is Lancaster Sound, which is where Uyarasuk grew up, particularly at Samirut on Bylot Island. The kayak has many baleen parts and may have been collected by John Ross in 1818. That in calm weather Uyarasuk went to sleep inside his kayak is the only recorded instance known to us for the Eastern Arctic. Far to the west in the Bering Sea, the Ukiuvagmiut of King Island sometimes went inside their kayaks, though not so much to sleep as to survive a capsize until rescued if unable to roll (Heath 1991:127).

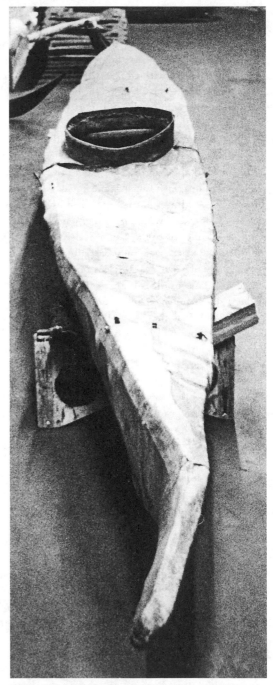

Figure 8.2—Kangitugaapik/Clyde Inlet kayak, collected between 1819 and 1820. Royal Albert Memorial Museum, Exeter. Unfortunately, this important early kayak is in poor condition. Courtesy John Brand.

Noah Piugaattuk (c. 1900–1995), the other prime informant recorded by the Igloolik Elders Project, confirmed Uyarasuk's report on bearded seal responses to harpooning; he also noted that seals go to the direction where struck so that in all likelihood they would go to the bottom of the kayak, thus the need to cast off the float immediately. Uyarasuk adds that when trying to kill the animal with the *anguvigaq*, or lance, the float could snag the kayak and cause a capsize. It was better to make the harpoon line longer, though not too long, so that the float would not get too close.

Piugaattuk also stated that boys were allowed to kayak only when they reached the age when they could look after themselves. Hunting training started out on the ice, where they were taught to get seals. With the wide, stable East Canadian Arctic design, practice from boyhood, as in Greenland, does not seem to have been necessary. Piugaattuk describes walrus and narwhal hunting in particular. His details on their behavior, including differing individual characteristics, are extraordinary.

> Among the various marine animals *aiviq*, walrus, is deemed most ferocious, so when a hunter is going to catch walrus they usually predetermine which animal they are going to harpoon. A hunter always pays close attention to the breath of walrus, so when there is more than one walrus they will observe the walruses that are in the water not aware of the presence of the hunters. After they have determined which of the walrus breaths appears to be better than the rest, they know which mean it will be much easier for them to catch. When this walrus comes up for air they will see that its breathing appears to be much smoother than the rest of the herd. They will see that the breath vapor is whiter than the rest as it is a fully-grown walrus. In addition the shoulder section is smoothed out. They know once a float has been attached to the animal it will not attack the float. When this walrus is pursued for the kill with an *anguvigaq* [lance] it will not attack the hunter. Instead it will try to flee rather than face the hunter when he approaches.
>
> Walruses in a herd will usually come up for air all at once. There will be some for which you will notice that the muzzle appears to be darker than on the rest and that the breath is short. As that is a fully grown walrus the colour of the skin is, of course, lighter. You will also notice that near the eyes the salt stains seem to be more evident, and when this walrus surfaces you will notice that the shoulder section appears to have sharp edges at the far back. This walrus

must be avoided. Once a float has been attached, it will attack the float. This type of a walrus is ferocious; it is not the type to be taken for granted though it may be full sized. The skin is darker than the rest. It may not be a fully grown walrus yet be large in size. These walruses will not hesitate to attack anything given the opportunity. When there are others around they are not chosen for the kill as they will face the pursuer and attack at once. They are deemed to be ferocious; this is also applicable to bears. Polar bears that are not fully grown usually act in the same manner. I was made to be aware of these signs as my father knew all about these things from his own personal experience, that is, to battle with walrus....

> It was important for us to know the internals of the large animals in order to disable an animal as quickly as possible, that is, for the species of animals that needed to be battled in the open water. This, of course, is applicable in moving ice when there is a build-up of pressure ridges close by where it would pose a certain amount of risk. Thus it is important to make the kill as quickly as possible with the *anguvigaq*....
>
> When walrus flee they will herd together. I started to get experience when we started to use a small boat while my uncle used a kayak, and he would harpoon a walrus. We were in a small boat and the only means of propulsion then was oars. It was up to us in our boat to make the kill with the *anguvigaq*.... I would lance the walrus... I was not strong but could throw the *anguvigaq* at a target having learned from boyhood. Since my family used to live alone, I grew up having to use the lance as soon as I was able to. Anything is possible if only one is given the opportunity.... So when my uncle harpooned a walrus the rest would go into battle with it. We had begun to make Kangirlukjuaq [Steensby Inlet] our home where we'd return after spending some time beyond it at certain seasons. We'd go much farther to the waters of Isuttuq. In summer there were more walrus in that area than others....
>
> When he harpooned a walrus the float would be taken completely under and soon resurface. Once it resurfaced the walruses would come up for air, this time all regrouped. So now they are in a herd as they flee, at the same time going to the one that the float is attached to. As the walrus flee the one which has been harpooned starts to fall behind the rest as the *niutaa*, the drag, slows down the animal. The *niutaa* is foaming white as the walrus pulls on it. The others *tutit-tuq*, lie close along side, which is to say they are helping the

harpooned walrus pull along the float and drag as they flee. Now the hunters go after them, but one should stay clear from the swirling of the water caused by the animals. This happens as the animals flee in that they swim closer to the surface and the animal which has a float attached to it is being helped by the others to pull. The harpooned walrus is not going to be left behind. The trick was to scare the animals away. My father told us to make sounds and bang on the boat so that the walrus that are keeping the harpooned one company would leave it behind. As we followed, the kayak was also in pursuit but to the side so as not to get directly above the animals....

As for narwhal hunting in a kayak, I spent only two summers with my uncle at Tununiq [North Baffin] where we each had a one. He harpooned a narwhal that was still not fully grown, and we started to battle. It fled from us, but soon my uncle made another attempt to fasten another harpoon line to it. But his paddle broke so he almost capsized. It was up to me to lance for the kill, so I went in pursuit. As it fled it was easy to keep up, but soon it dove deep into the water taking the float down with it. Soon the float surfaced straight up and did not move: this was an indication that the whale was not moving. The harpoon line just went deep, straight down in the water. My uncle went right up to the float and kept an eye on the harpoon line. Soon he said that it was not going anywhere. As it turned out, narwhals that do not drag the float are much harder for making a quick kill. On the other hand with a narwhal that has been harpooned and flees you can keep close to the float as the narwhal pulls it in the same general direction. As soon as the narwhal starts to come up you can paddle up quickly before it surfaces for air, and as soon as it surfaces you can attach another harpoon line. But this narwhal had taken air too soon so he [Piugaattuk's uncle] was forced to rush to it, breaking his paddle which almost made him capsize.

He went by the float and stood in his kayak to keep track of the harpoon line which ran straight down. The narwhal did not move but stayed stationary. He said the narwhal had taken in air fast. As it turned out when they do that they are hard to catch. When they do not move or flee away from the hunters, it is difficult to determine in which direction the narwhal would be surfacing to take in air.... He kept watch of the harpoon line, then suddenly said to turn in a certain direction, so at once without hesitation I turned my kayak.... The harpoon line started to move to the direction

where the narwhal would be surfacing so it was that way he asked me to face. Once I turned to that direction he said the narwhal was coming up for air. Sure enough the narwhal surfaced right in front of me. The second time it surfaced, as it dove I harpooned it. Once two floats had been attached to the animal this kept the animal from going a long distance. In addition should one of the heads pull out of the skin there would be the other still attached. At this time he told me to lance the animal for the kill. The harpoon lines continued to hang straight down, but once they started to move he would tell where to go to where it would surface for air, so at once I would go to that direction to lance the animal. So when the animal surfaced in front of me I lanced it, but as I had not been told where to lance a narwhal prior to this, it dove again. He said to me that I had lanced it a little short: I had aimed where the intestines connected to the rest of the innards. As it was not a walrus he thought that I lanced too short but it was almost right on. I had never seen a narwhal lanced before so I just aimed at the spot where I would have aimed for walrus (Piugaattuk 1990, IE-136, translator Tapardjuk).

The uncle's near capsize and broken paddle resulted from not being able to dodge the float, which caught the stern of the kayak. To stay upright, Piugaattuk's uncle plunged the good part of the paddle in the water, then tied the parts together. Piugaattuk's uncle made kayaks that did not capsize easily since he often went out alone. Kayaks that were narrow, at least in Eastern Arctic terms, were silent as well as easy to paddle but risky for hunting alone. Lack of rolling techniques discouraged lone sorties in narrow kayaks, in contrast to the Greenlandic practice. For the purposes of design study, it is important to know that in North Baffin, as elsewhere in the Eastern Arctic, kayaks are thought to come in two widths: narrow and wide.

My uncle and myself when we resided at Tununiq used these kinds of kayak: I had one that was made to get close to an animal so the frame was narrower while length was about the same. It was very easy to paddle in and there was hardly any sound of cutting through the water. My uncle's kayak was wider and harder to paddle. When there were two of you on different kayaks it would appear as if the other was paddling so effortlessly, especially when one was trying to catch up or to keep at the same pace as the one in front (Piugaattuk 1990, IE-136, translator Tapardjuk).

Arviq or *aqviq*, the bowhead or Greenland whale (*Balaena mysti-cetus*), was by far the largest animal hunted by arctic peoples, measuring up to 60 to 65 feet (18 to 20 m) long and weighing between 75 and 100 tons (68,000 to 90,000 kg). This great polar whale is usually associated with *umiak* hunting, but was also hunted using the kayak. In historic times in North Baffin and Foxe Basin, the open skin boat was rare, perhaps due to a shortage of wood needed to construct its framework. The kayak was thus the primary whaling vessel. Thanks to Inuit cultural memory, some technical details are available regarding this crowning achievement of Eastern Arctic kayaking. In 1973, at Mittimatalik, Guy Mary-Rousselière gathered the following information about North Baffin whaling:

> If a Greenland whale was sighted, a kayaker, even alone, did not hesitate to chase it and to harpoon it. Usually, when a hunter got his harpoon into a whale, other kayakers in the vicinity would come to harpoon it and help dispatch it. But it is said that a man alone could sometimes manage to get a whale. Nutaraajjuk, Qumangapik's father, killed a whale single-handed at Igaajjuaq, near the present settlement of Pond Inlet. The lance was used for the killing; but this was not always sufficient, and sometimes the bone tip of the paddle was used to enlarge the wound and reach the heart. For that purpose, Inuguk's father and a few others had metal tips on their paddles. There was a certain amount of danger involved in hunting the big whale, but it is said that the animal was not exceedingly difficult to kill. A skin float of *natsiaviniq* (young seal) was used for Greenland whales. Sometimes two or three of these were carried on the kayak. The float has to be thrown away very quickly after the whale was harpooned, to prevent the kayak from being upset, because the Greenland whale—as well as the narwhal—when hit dives vertically (while the walrus dives more obliquely). The towing of the dead whale to the nearest shore was a time-consuming undertaking, even when several kayaks were pulling (Mary-Rousselière 1991:62).

The specification of a float of young ringed seal for the bowhead was probably because a small float had less chance of dislodging the harpoon head, particularly at the first hard dive. There is also the drag to consider. The extra-large whaling harpoon would not be used when hunting from a kayak, so the harpoon head was much smaller than the special whaling one, which measured about 10 inches (25 cm) long. In contrast, whaling from an *umiak* with the larger

harpoon used multiple larger floats as well as the drag, or tambourinelike hoop of skin held perpendicular to the pull of the harpoon line. With kayak whaling, repeated harpooning to attach more floats on separate lines was common, with companions assisting after the first strike. Occasionally, a hunter got a whale singlehandedly, like Nutaraajjuk—quite a feat. Of the others named above, Qumangapik was the main builder of the 1973 Mittimatalik (or Pond Inlet) kayak reconstruction; Inuguk was an old informant (Mary-Rousselière 1991:43).

Piugaattuk told of whaling from two canoes at Kapuiviit around 1961. He provided details of where to strike a bowhead whale:

> I told my companions that we all are familiar with narwhal in which the back sinew has an area where the *taqamuk* [fascia] is greater where the blubber connects with the meat. I told them it might be the same with bowhead whales, that this part might be strong enough to keep the harpoon head from drawing after it was attached. Should a harpoon draw, then the area around the mouth is the place to aim for by what I have heard using an ordinary harpoon head. When you have a choice of where to strike, then you can aim for the nostril. Once the harpoon head is attached to any of these areas, it is going to hold. So then when me and my son Maliki got to a bowhead I had everybody informed that they should go for the back sinew area when harpooning. Should the harpoon head draw then they could aim for the corner of the mouth. My harpooner was in front of me but I examined the whale and discovered where the flukes were. I harpooned the whale to the sinew; it did not budge. My partner also struck on the back. We both threw the floats to the water and got away from the bowhead. When it started to dive we moved away toward the sea where we went back and forth then soon stopped. We saw it moving toward the land.
>
> Once we knew it was going to go landward I placed myself in front to make the kill. When we got to it the first time after the harpoons were attached, I shot it trying to make the bullet penetrate the skin and go as deep as possible. Our intestines have an attachment near the kidney; this goes for all of the game animals. In the case of walrus in the water when this attachment is severed the walrus cannot live long afterwards. When we got to the bowhead I shot it through that area; I shot it a few times and noticed that it was slowing down. When we got to it once more with an *anguvigaq* lance... I examined the whale, how it was facing, and found out where the foreflipper was. Once I got the

location of the foreflipper I knew and pictured where the heart was. When I plunged in the lance, almost the whole length of it, the whale convulsed meaning that the lance had struck home in the heart. (Piugaattuk 1990 1E-136, translator Tapardjuk).

The target areas for harpooning and wounding to weaken or kill are the same in kayak whaling; however, it is more difficult to lance the heart from a sitting position. Use of the sharpened paddle may have been necessary to produce a wound to the heart. Seeing where to strike may also have been more difficult. The *taqamuk*, mentioned above, is located on the back of the animal's head, an area that is accessible as the whale surfaces to breathe. It's a good target because the fibrous muscle sheath provides a strong anchor for a harpoon head. The hunt may have been more arduous from a kayak, even with more than one kayaker participating. In the past, the *umiak* provided the advantages of harpooning and lancing from a standing position—part of the reason why it was preferred when available. Furthermore, a crew consisting of several men provided additional power and task specialization.

Towing the whale to shore was a hard task even with a number of paddlers in kayaks or *umiat*. The whale's huge mouth was tied shut if necessary to keep the whale from filling with water. Landed at high tide, the big animal would be butchered and shared. Piugaattuk heard that in the northwest Hudson Bay, where imported whaleboats were used, the harpooner who made the first strike, the pilot, and the sail handler got the biggest shares. The handler and the harpoon-line tender each measured a section the length of an armspan. With kayaks, the order of harpoon strikes determined the ordering of shares. The back sinew area was so wide that the meat had to be cut into short pieces. The weight of the cache squeezed the blood out of the meat at the bottom. Meat was also dried, usually from the back sinew area.

Native hunters captured a small number of whales relative to the formerly large population, certainly far less than the numbers slaughtered by white whalers of the 19th century. For the Inuit, even a single whale every year or two was of great significance, providing a mountain of meat and blubber equal to scores of seals, walrus, narwhal, and beluga. Whales provided not only food and fuel but also bone for house frames (Lee and Reinhardt 2003) and sled parts, sinew for sewing and tying, and baleen for water-resistant kayak lashings, deck fittings, and drag hoops. Whether by kayak or *umiak*, whaling enhanced the quality of arctic life (Figure 8.3).

KAYAK HUNTING ON LAKES AND RIVERS

An inland kayak for lancing swimming caribou in rivers and lakes at crossing places has been distinguished at times for the East Canadian Arctic. The notable instance is the distinction made for northwest Hudson Bay kayaks among the Aivilingimiut, the southern division of the Igloolik (Iglulik) Inuit. This distinction can be illustrated through two contrasting models collected by Captain George Comer (Boas 1901–1907:77, figs. 106a, 106b). These models, which measure about 2 feet (75 cm and 80 cm) long, are fairly detailed, although profile views alone make it difficult to ascertain vital aspects such as hull cross section and plan outline. The inland lake model (fig. 106a) is a continuation of the old narrow multichine of Foxe Basin and West Hudson Bay with long, idiosyncratic end horns of the sort recorded in the 1820s at Igloolik by Parry and Lyon (Lyon 1824:321–322; Parry 1824, vol. I:506, plates facing pp. 274, 358, 508) and seen in 1847 at Naujan/Repulse Bay by Rae (1850:94). This feature was retained into the 20th century by the Caribou Inuit from the Thelon River south to Churchill.

In 1994, John Heath saw a British Museum specimen without provenance information that had emerged after several decades in a box. Based on his photos, the kayak can be identified as the old Foxe Basin type with the bow horn apparently flowing straight out of the hull sheer, angled up without jog or bend to point it lower. Since Comer's Aivilingmiut lake model and Caribou kayaks have lowered

Figure 8.3—White whale hunter, possibly Qilalugasiurvik, c. 1865, southeast Hudson Bay. *Avataq*, or sealskin float, in foreground. The prow appears to be intermediate to the West Hudson Bay design. Photographer George Simpson McTavish. Courtesy National Archives of Canada, no. C-022941.

bow horns, the former Foxe Basin design appears to be a variant cousin. Other distinguishing features include strong longitudinal rocker and heavier framing.

The Aivilingmiut sea model (Boas 1901–1907: fig. 106b) resembles an East Canadian Arctic design but retains some of the old multichine's features. As in the inland model, the cockpit hoop is round and tilted up in front for easy access and wave fending. The main deck cross lines are not a parallel pair but are joined in the middle and may be two loops from the gunwales linked by a tightening line. At the stern is a relatively prominent upturned projection, a short version of the West Hudson Bay angled horn. The cockpit front is amidships and since the hoop is not long, the sitting location is not markedly rear-set as in the East Canadian Arctic kayak. While the afterhull is full, the maximum width appears more forward. The afterhull is not significantly shallower, and the bow is not especially deep at the forefoot, nor high at the tip. The prow is narrowed with a sharp cutwater, which is raked but does not have the usual impressive stretch due to the low bow. The bottom cross section is presumably multichine with a flattening underneath, assuming beam is greater than in the inland form. Hybridization between the Foxe Basin–West Hudson Bay and East Canadian Arctic designs is evident; nevertheless, the sea model contrasts with the inland one so that, at least for the late 19th century, two kinds of Aivilingmiut kayak may be recognized.

The low bow of a 1967 Aivilingmiut reconstruction from Nauyaat in the Canadian Museum of Civilization (IV-B-4127) resembles that of the sea model; however, the cockpit hoop is an elongated eastern arctic form. A bulky-looking kayak, its most striking feature is an aft-set cockpit, due to the afterhull being a couple of feet shorter than would be expected with the long bow. The forefoot curve projects a little below for a skeg effect as in North Baffin. The bottom is well rounded due to numerous strips set close together on shaped board formers, rather than ribs. The middle of the bottom is flattened and the gunwale planks are vertical.

Framework construction is highly modified since the craft was built for sale to a museum, commissioned by David Damas for the Human History Branch of the former National Museum of Canada, now the Canadian Museum of Civilization. He says it was made by Tagurnaq, a man in his 40s, advised by Kiyuaqjuk. At the time, Aivilik kayaks had not existed for years. However acculturated in construction, the specimen is useful for study, showing that the design in the region has a rounded bottom with a flattening underneath,

whereas with the classic East Canadian Arctic type, the design is flat-bottomed with flared sides.

According to Bernard Saladin, an inland adaptation like that of northwest Hudson Bay existed on the eastern side of the bay. A small, light example skinned in caribou was reconstructed for him in 1968 by Tuumasikallak at Kangiqsuk or Payne Bay on West Ungava Bay (CMC IV-B-1445) and is only 14 feet 9 inches (450 cm) long and weighs 45 pounds (20 kg). A cockpit hoop tumpline helps portaging. The shape is East Canadian Arctic, with the flat bottom and hard chine, or edge where the bottom meets the sides.

The short kayak of central east Hudson Bay of the Inujjuaq–Puvirnituuq region, though sealskin-covered and not as small and light as the Kangiqsuk reconstruction, is designed for the river and lake chains running inland, as well as seacoast use. Its comparatively small size was for ease of portaging, unlike the large, heavy type used to the north. In southeast Hudson Bay, the kayaks used for the big rivers reaching inland, such as the Great Whale River, are bigger, measuring nine ringed sealskins long. Aside from the Aivilingmiut case, where different kayak designs overlapped in a mixed late 19th-century situation created by commercial whaling, there seems to have been no distinct inland model in the East Canadian Arctic area. Bottom form alone, whether flat or multichine, does not distinguish a kayak as a sea or inland model. For the north Foxe Basin caribou hunting on Baffinland, informants note no difference in form.

Any description of Baffinland caribou hunting has wider ethnological importance given the poor knowledge of the region. The extended passage quoted below by Piugaattuk begins with a discussion of seal and walrus hunting as complementary parts in the seasonal round. The kayak hunting of caribou is set into a fuller economic and ecological context. Molting flightless *kanguq*, or snow geese, first the adults and then the fledglings, had been driven into stone weirs at Ipiutit before the following stage:

> When they [Baffinlanders] were planning to set up a summer camp at Inuksulik, the spring would be spent hunting walrus at Qikiqtaarjuk. At the southern part of Kingaqjuaq near Qikiqtaaluk there is a small island known as Qikiqtaarjuk. There they would spend the summer. Before the ice melts causing bodies of water, they would hunt seals through breathing holes in order to collect blubber for fuel. The seals in that area do not move around as much, whereby the breathing holes for each are not interconnected with others as it is around here [Igloolik area]. In the waters around Ikpiit

the breathing holes on the ice are usually separated from each other so the seals do not move on to the next set of holes. When they see more than one seal basking on ice, they would approach and start to hunt through the breathing holes. When they caught plenty they would have a sufficient supply of blubber for fuel. They caught as much seal as they could so they would concentrate their time in the summer hunting caribou. That's the way it was at Baffin Island. In summer they depended solely on *qayaq* for sea travel. The younger people were given the task of hunting with arrows.

At Qikiqtaarjuk when the ice started to melt and it was no longer suitable for hunting seals through breathing holes, they started to hunt walrus so they could stockpile them in caches. The older men would leave by kayak. Irloo was a boy at the time. He could see kayaks approaching, tied together. They would be coming toward the camp for a long time and finally arrive. They would be towing walrus already cut up and fixed in a way that it would be easy to tow them using their harpoon floats. The *qayaqs* were weighted inside as well on top so they would be ballasted. By helping each other they would finally reach the shore. It is certainly tiring. Once they made the caches when the ice was melting to the point of rotting, they would move the camp farther up from Qikiqtaarjuk. There were three elders in that camp with all the youngsters. One of them, Toomigiit, did not have many children; he had married a daughter of Tapaattiaq. They did not have many children; as a matter of fact, they did not have any sons. Therefore he lived amongst his father-in-law and his peers. They had another son and a number of daughters. Aksakjuk was one of them and he had two or three sons.

They wanted to hunt caribou when they were crossing the lakes at Inuksulik during the summer and autumn. So they moved there when the ice was suitable to travel on, passing Kingaqjuaq and going beyond to the waters of Inuksulik. There they would spend the summer. The elders would go to the lake known as Inuksulik. When they reached their destination they made camp. Once they settled in, their younger people would move inland and hunt caribou using bow and arrows. They would make caches as they continued to hunt caribou. They did not have any dogs with them. While the younger ones were out on the land hunting caribou, the elders had settled on the shores of this lake. They were able to catch some caribou who had come down to the lake and swam across it. There the elders would stay for the summer. Tapaattiaq was not very big and was not popular,

but he worked hard to keep supplies. He used to live in a temporary camp alone with his wife out on the mainland. He was very active.

When autumn was approaching the elders would wait for their return from inland. When the bull caribou's antler started moulting, towards autumn when temperatures were getting colder, the caribou that had been migrating towards Kangirlukjuaq would slow down and they would start facing in the other direction towards Nattilik and Nalluarjuk. The caribou would start facing that direction. At this time the ground started to frost. The caribou would begin to migrate to the opposite direction when the ground started to frost at night. The people who had been hunting inland headed back to where they had left their elders on the shores of the lake when the caribou started to migrate in the opposite direction. This information was passed on from Tapaattiaq; he would go back to the elders when the caribou had started to migrate. When a large herd started to cross the lake the hunters would go after them in their kayaks. The caribou that escaped them would land on the shores, and the hunters with bow and arrows would wait where the caribou would land so they could catch them. Then they joined the people on the shores of the lake. They had to work hard because of inadequate hunting tools; this particular hunt was for them to prepare for the winter. The fuel supply prepared the preceding spring would be the only caches that were of marine animals, *mingukturniq*....

That's the way it was at that lake known as Inuksulik. The lake is positioned across the caribou path and there is a narrow spot where they cross. The men would take to the water with their kayaks when the caribou has swum halfway. They would approach them head on, and the caribou would start to swim away from them. The kayaks would position themselves so that the caribou would be between them. One of them was left handed and the rest right handed. The one who was left handed would be positioned at the right side of the herd, and his companion would be on the left, the rest following from behind. The two who had positioned themselves alongside the herd would start to move in closer so the caribou would move together. Meanwhile the kayaks that had been following behind started to move up to the caribou. As they approached with strong strokes of their paddles they would move between the caribou and with the paddle they would anchor themselves by pressing down one end to the back of the last caribou and the other to the kayak. In this position the man would choose the caribou farthest ahead of him to stab it with the lance. He would keep lancing the caribou that were behind the herd, working his way from the farthest to the closest. The caribou that he was anchored to would be the last one to be lanced. The person who was positioned on the right side would be paddling in tight alongside the caribou. He would anchor himself to the closest caribou by pressing the ends of the paddle to the back of the caribou and his kayak. The other hunter at the opposite side would also do the same. They would face the caribou until they had killed them all. Then they would go after the caribou that might have left the herd and repeat the same thing they had just done; that was how they were able to get plenty of caribou for food. Thus they hunted caribou at Inuksulik. The younger men stood by on the shore in case a caribou would reach it and climb up. They were prepared to intercept any caribou that might land. If a caribou set foot to the ground and started to flee it would be shot with an arrow. Such was the way they hunted in order to get a sufficient supply for the winter at Inuksulik. After they caught all the caribou they started towing them towards the camp, going back and forth as they could not tow large numbers at once. They would leave the carcasses on the dry ground to drip off. A large area would be covered with dead caribou.... With all these on hand they began to determine who would get caribou; they began looking around to give each hunter his share and started to separate them. The elders determined who got what. When there were bull caribous each would be given one. They started to get their share equally distributed so each of the hunters could make a food cache. The caribou would be gathered into separate bunches. They would pull them to higher ground and scrape off the water at the same time. Then they started to skin and cut them up after they had settled the distribution; that's the way it was done at Inuksulik (Piugaattuk 1991, translator Tapardjuk).

The elegant lancing technique using the energy of the last caribou to keep up with the herd seems practical only with a comparatively stable kayak, which is perhaps why it is not recorded from Caribou, Nattilik (Netsilik), or Copper Inuit with their tippier designs. The mention in the early part of the passage to towing walrus "already cut up and fixed" is explained elsewhere as the deboned meat being laced up in skin in rounded packages. Floats buoyed up the heavy pieces. To tow them, the kayaks were ballasted with load for more momen-

tum, as was done for other big catches such as whales. Making the kayak heavier might have slowed it down but gave a steadier pull.

For speed, Piugaattuk mentions tilting the blades, a detail of interest to recreationalists:

> When kayaks are in pursuit of swimming caribou they must go as fast as possible; this is called *upaksattut*. The hunters would tilt their paddles so the blade is angled in order that the kayak slices through the water more smoothly. Otherwise, if you have your paddle blade down you will see that the kayak has a tendency to poke into the water at the front end making it harder to paddle because of the drag the paddle will cause, in which manner one can be certain that the hunter will have difficulty keeping abreast with the others who would be making good time. When the paddle is tilted the kayak will not have a tendency to dig into the water but will run on top (Piugaattuk 1991).

Blade tilting for speed is done in Greenland, too. No doubt it is a widespread Native technique.

The caribou hunt described above entailed many kayaks, but it may have been more common to employ two kayaks. Piugaattuk's account below is significant not only for the technical details but also in confirming the commonality in caribou hunting practices over a wide area, since the same things are done by the Caribou Inuit of West Hudson Bay. Historically, the Caribou and Igloolik Inuit are known to have had the same kayak design, with minor variation, in the 19th century. Even today some Igloolik Inuit show more of an orientation toward their southern neighbors than to the Tununirmiut and Tununirusirmiut of North Baffin. Part of the relatedness may arise from closer contact in the later 19th century due to West Hudson Bay commercial whaling.

Here is Piugaattuk's description of the two-kayak method for intercepting caribou:

> As soon as the eider ducklings start to go to sea is the time when the thickness of the caribou fur is prime for clothing material. At this time they would start for the inland, usually to a lake where the caribou would cross, *nallusiuttuq*.... The people that are going to spend the summer inland would be the hunter with his wife and another couple. There were usually two kayaks taken so that they could complement each other when they go after the swimming caribou in the lake.

Once they reach their destination of a *nalluq* or crossing, they call this place their home. They make camp across from the side of the lake where the caribou take to the water as they start to cross. Along the route that leads to the lake *inuksuit*, man-like stone pillars, will be erected so that the caribou will not make detours. A line of *inuksuit* would be erected, and a thong would be run along them with pieces of skin or other material tied in between the cairns so that they could be moved around by pulling on the thong. There are two lines of *inuksugait* that narrow down to the lake. As the caribou get to a line of *inuksugait* they will follow it down from fear that they might be humans. When caribou get spooked they start to run forward, but when they come across something that is moving they will flee to the lake. Once they started to swim across the lake a kayak would go after the caribou in the water.

As the hair on the caribou starts to get thicker the caribou will start to migrate in the other direction. Sometimes when caribou get to the lakeshore they sometimes will back off. This usually happens in early autumn which sometimes makes it difficult for the people to get caribou for food, especially when there are quite a few people in the camp. The meat is soon gone when an animal is caught using only the bow and arrow. They could see the caribou on the other side of the lake when they would back off before going into the water and return to the hills where they would start to feed. Because the caribou are no longer swimming the lake, the hunters are running low on food.

Soon they decide that one man must go out and herd them. The man will cross the lake in a kayak while the other will wait at the camp. Once the hunter crosses the lake, taking a *pukiq* or white furred belly skin with him, he makes a detour after pulling up his kayak. As he walks towards them he must pass the herd of caribou so that they stand between him and the lake. After he is in position he will attract the attention of the caribou by flashing the *pukiq* so that the white fur is seen. The best way to achieve this was for the hunter to strike the ground with it. The purpose in doing that was to scare the caribous and get them to flee. Whipping the ground with the *pukiq* makes a loud noise so the caribou start to flee; he runs after them. As he follows them he tries hard to spook the caribou so that they continue to flee and get close to the lakeshore. The caribou may take to the water out of view of the hunters, so he runs as fast as he can to a rise to keep track of them. Once he gets onto the rise he will see that the caribou have started to swim across the lake. He at once starts to *aulaugajuq*, that is, facing sideways he bows back and forth, telling the people in the camp that the caribou had taken to the water and started to cross the lake. After he catches the attention of the people, he dashes to his kayak to go after the swimming caribou.

When he gets in his kayak he goes after the caribou. The hunter from the camp has gone towards them from the other side of the lake. The man that had driven the caribou comes from the opposite direction so that he can meet the caribou that turn back. Soon the caribou are between the two hunters. They herd them away from the land so that the caribou will not be able to land quickly. Soon they get among the caribou and start to lance them....

After the hunters have secured food they wait longer well into the autumn. They occasionally go out on hunts away from the migration route. They are very careful not to go to the migration route; otherwise they might disrupt it.... When a hunter spots a caribou he tries to get as close as he can in order to shoot it with his bow and arrow. When a caribou flees from the hunter it flees away from the wind. Should it be a bull the hunter will go after it. After a while he keeps it in his view so that he can continue to pursue it, which will in all likelihood take him all day while there is still some light. As he goes after it the caribou will eventually slow down. When the caribou starts to walk the hunter continues following but keeps his distance. This process is called *narjungniattuq*. Once he gets ahead of the caribou he waits for it to come to him. When it gets close enough he will shoot. After the people in the camp are resupplied with food, they continue to wait in their camp for the migration. As the autumn progresses they will spot herds migrating on the other side of the lake (Piugaattuk 1986, IE-202, translator Tapardjuk).

The caribou lance is thinner and longer than that for sea hunting, with a head made from antler. The Amitturmiut (Iglulingmiut) term is *ipuligaq* and to the south is *iputuyuq*, presumably in Aivilingmiut. Caribou Inuit call it simply *ipu* (shaft) or *kapuurauyaq* (piercer). Lances have particular meaning because, along with the bow and arrow, they were expressly killing weapons. Harpoons and darts could kill, too, but their primary purpose was game capture. The kayaker and his equipment functioned together as a highly developed hunting machine, key to the success of Inuit adaptations to the Arctic.

East Canadian Arctic kayaks are part of the greater sociocultural continuity in the region. Their variation, with a couple of exceptions, is not on the order of that among the West Canadian Arctic kayaks of the Caribou, Nattilik (Netsilik), and Copper Inuit. The major exception is the distinctive North Baffin–North Hudson Bay kayak, which may be a hybrid or transitional form in a border region between two different kayak families, the flat or shallow V-bottom East Canadian Arctic–Greenland and the multichine West Canadian Arctic–North Alaska. "North Baffin" is loosely defined to include what might be called "East," for example, Kangiqtugaapik/Clyde Inlet, and "West," for example, Agu Bay. "North Hudson Bay" includes the Foxe Basin (Map 5) as well as the bay proper along Roes Welcome Sound and around Southampton (Sadliq) Island, where the Sallirmiut kayak was said to have been similar to the sea kayak of the Aivilingmiut (Mathiassen 1927, vol. I:276), that is, similar to the North Baffin multichine Eastern Arctic variant. Since the "North Baffin–North Hudson Bay" label is unwieldy, it will be shortened to "North Baffin," thus giving primacy to the original home region.

The North Baffin kayak looks like an East Canadian Arctic design that has taken on certain West Canadian Arctic features, or vice versa. Besides the shallowly rounded multichine hull with light curved ribs, the North Baffin kayak has a couple of smaller details reflecting West Canadian Arctic design influence. The primary deck line before the cockpit has the Caribou kayak configuration with two branches at each side converging to a single middle section. The end block in the stern assembly recalls the Nattilik (Netsilik) framework of the region to the west around Boothia Peninsula (Map 5). In a 1930s Clyde River example, the stern has a jogged projection recalling the Nattilik model again (Brand 1984:23). Sometimes there is a drain hole on top of the stern block plugged by a stick, halved lengthwise, with the flat face resting against the block—the same system found in the Caribou kayak against its stern horn. That these minor nonperformance traits are western Canadian Arctic while overall form is eastern suggests that it was the western design that became easternized, retaining the less cranky multichine cross section.

North Baffin sterns, at least the Tununirmiut and Amitturmiut (Iglulingmiut) ones from the 20th century, are characteristically plain and low without a handhold projection. Maximum width is usually amidships instead of aft, but otherwise the top outline is East Canadian Arctic, the afterhull broad, the bow long and narrow. A turn-of-the-century model representing the Aivilingmiut sea kayak is basically East Canadian Arctic in hull form with a large cutwater bow, but has a small round cockpit hoop that looks West Canadian Arctic and a water-bird-shaped paddle rest strung on two cockpit front deck lines. The model's stern is upturned and attenuated in a small projection recalling the old Igloolik sterns illustrated by Lyon (Boas 1901–1907:77, fig. 106b; American Museum of Natural History catalog no. 60/2838a).

Particular to North Baffin and not derived from West Canadian Arctic models are a blunt deep stem tip in profile and an S-shaped cutwater outline below for a so-called Clipper bow. In a previous study (Arima 1987:100), the blunt stem appeared to be due to recent gunwale construction using a board ripped in half, on edge, save for the first foot as described by Father Mary-Rousselière (1991); however, the Kangiqtugaapik or Clyde Inlet kayak collected during Captain William Parry's first Northwest Passage voyage (1819–1820) has the feature, albeit in smaller form. In this earlier North Baffin type, the gunwale planks are not sawn but hewn. Recently, the cutwater strip joined flat against the bottom of the squared-off front end of the gunwales takes on a noticeable S shape (or Clipper bow) as it first bends away concave to the outer forward side and then recurves convexly below, forming a downward-bulging forefoot. This last feature is effectively a tracking skeg, noticed on the Uyarasuk model that was revived at Igloolik in the 1980s for recreational kayaking. It is about 17 feet (518 cm) long and 25 inches (64 cm) wide. The Kangiqtugaapik kayak collected during Parry's voyage is much the same size, 16 feet 9 ½ inches (512 cm) by 26 inches (22 cm), as tentatively reconstructed on paper. Most known North Baffin examples are wider and larger, up to about 22 feet (7 m) long. The Kangiqtugaapik kayak could have been at least a half inch wider, beamier since the sides are collapsed inward from the cockpit aft to the quarters, probably from the pull of the cover. If so, it was designed to be extra stable, even during harpoon-hunting times.

Since the shallowly multichine wide hull with a slightly V-shaped bottom provides less lateral catch in the water than the flat-bottomed and hard-chined section standard in the East Canadian Arctic design, the North Baffin variant would not track as well without the skeg forefoot. Uyarasuk mentioned the need for edges fore and aft when asked if the kayak tended to turn at speed, as follows:

> Yes, but you can control that without difficulty especially
> when you are going after an animal always trying to keep it

to your right. The stern should have an edge that will keep the *qayaq* going straight…when it does not have one it tends to turn too easily. That keel at the front must have edges where it is submerged. When one is going fast in a *qayaq* this will keep it going straight; when it does not stand out the *qayaq* will turn too easily (Uyarasuk 1991, interviewer and translator Tapardjuk).

Another reason for having a well-developed, downward-bulging forefoot was silence. As Uyarasuk explained, "If it was away from the water, where there were some waves around it would make noise with the waves banging against it, making it impossible to get close to the animals." In a probably independent development, the West Coast, or Nootka, canoe has its skeg effect cutwater bottoms enlarged downward in the sealing model for silence. At about 17 feet (518 cm) long, Uyarasuk's design seems small for North Baffin:

> When I made the *qayaq* I tried to design it to suit myself. I always tried to make it so that it is not too short; it would have to be able to carry two additional persons. This was especially handy when one had to use it to get to the land from the ice when the ice was getting bad before the break up (Uyarasuk 1991, interviewer and translator Tapardjuk).

One of the passengers would likely lie inside, probably forward on the back with the head to the cockpit. Then the other could be behind the kayaker on the deck either prone or sitting, or inside if it was deep enough aft. The kayak and the sled were used together for transportation when hunting out on the ice. The kayak carried all the gear on the sled from the ice to the land. The kayak was placed on the sled upside down, stern forward.

Uyarasuk mentioned another consideration regarding cockpit opening size:

> As for the *paa*, one tried to make it large enough so that you could jump into the *qayaq* when the wind was blowing in line with the shore. When the waves are too big for stability all you have to do is jump into the *paa* to get in. The people before us, whom I was fortunate to have seen, had numerous *qayaq* and when the waves were too big at the beach they just put the *qayaq* into the water and jumped in (Uyarasuk 1991, interviewer and translator Tapardjuk).

The cockpit was located farther aft around Igloolik for quicker turning to evade aggressive walrus (Piugaattuk 1991). On the North

Baffin cross section, which in comparative study is considered to be multichine, it may be surprising to hear that the design objective was not to make the main body of the hull more rounded, but rather to have the bottom "flat." As Piugaattuk (1991) explains:

> When the *qayait* were being built they made certain that the bottom is not rounded. When rounded the bottom part that is going to be submerged might pose a certain amount of danger. When the *qayaq* tilts it will not stop and can upset too easily. It used to be said, "*Qilausingatamiguuq nangiannammat*," "Since it is shaped like a drum, it is deemed dangerous."
>
> A *qayaq* built in this shape is also dangerous when it is going at a fast speed. Should one force it to go fast, once it starts to tilt it will continue to do so until it capsizes. So it was important in making a *qayaq* that the bottom be made flat at the aft section where the width is widest (Piugaattuk 1991).

The Tununirmiut paddle has pointed side projections outside the grips to allow water running in from the blades to drip off. The shaft can be cut for such points in outline or an antler point may be fastened on. The multichine East Canadian Arctic kayak is found from the west end of North Baffin to Kangiqjuaapik or Clyde Inlet, where there is an overlap between the Tununirmiut, also called Aggumiut or Northerners, and the Akudnirmiut, who extend southeast to Cumberland Sound. The flat-bottomed East Canadian Arctic kayak proper begins to occur among the Akudnirmiut.

FLAT BOTTOMS The classic flat-bottomed East Canadian Arctic kayak existed in Baffinland southeast of Clyde Inlet and the Labrador Peninsula, retaining the same shape over a very extensive area. Although a flat bottom is the norm (Figure 8.4), sometimes it is slightly V shaped. When the bilge stringers, or longitudinal lathes, are grooved to seat the ribs, as they often are, they usually angle up slightly, giving a bottom contour that releases water pressure to the sides, an effect that is increased when the stringers are wide. There can be some significant variations in the flat bottom. The primary variation, however, is size, which entailed other adjustments in form. There were also differences resulting from local conditions, materials, uses, and stylistic trends. A general tendency in historical times, as noted above, seems to be a progressive increase in size over three or four centuries, probably stemming from the increasing supply of iron tools

Figure 8.4—Flat bottomed kayaks on stands, Kuujjuaraapik, 1949. This photo is a record of the East Canadian Arctic bottom form in East Hudson Bay. Photographer S. J. Bailey. Courtesy National Archives of Canada, no. PA-110863.

and weapons. By the 19th century, the East Canadian Arctic kayak became larger, between 20 and 27 feet (6 to 8.2 m) long. Kayaks remained small for portability in East Hudson Bay, inland Ungava, and East Baffin.

Franz Boas writes that, in the 1880s, the Baffinland kayak, excluding North Baffin, "is from twenty-five to twenty-seven feet [7.6 to 8.2 m] long and weighs from eighty to one hundred pounds" [36.3 to 45.4 kg] (Boas 1901–1907:408). These length figures are the maximum and may have been the ideal when builders had the means to take the East Arctic design as far as they wished. Firearms and ammunition in adequate supply for a developing economy and trade lumber combined with aspirations to build bigger and better. Two impressive 27-foot (8.4 m) examples, evidently from Atlantic Labrador, survive in Canada in the Newfoundland Museum (NF 4177) and Royal Ontario Museum. Boas had shorter Baffinland specimens collected for him in the late 19th century: a 19-foot-8-inch (6 m) example from Cumberland Sound and a 21-foot (6.4 m) from Savage Island, with maximum widths of 22 inches (56 cm) and 24 inches (62 cm), respectively (Boas 1901–1907:13). In the early 1880s, Turner gave 18 to 26 feet (5.5 to 7.9 m) for length for the Ungava Bay region and added that the maximum width is a third of the distance aft of the cockpit (Turner 1894:237). If the beam is so far aft of the cockpit in part because the cockpit was more forward in the 19th century, then the Ungava Bay kayak may have been closer in design to the Atlantic Labrador model.

There were minor differences in kayak form among different Native groups. Among Baffin kayaks, for instance, the Frobisher Bay model was built more shallowly than neighboring ones, presumably for less windage. At Ivuyivik, Ainalik explained that he preferred 24-foot (7.3 m) to 23-foot (7 m) kayaks in order to avoid swerving off course while coasting in on a seal to harpoon; the greater length was awkward. In the Puvirnituuq–Inujjuaq region of East Hudson Bay, length was kept smaller and lighter, at about 16 to 17 feet (4.9 to 5.2 m), because it was necessary to portage on the river and lake chains running inland. Farther south around Little Whale and Great Whale rivers, the inland kayaks were longer, measuring about nine ringed seals in length. Perhaps because the Puvirnituuq–Inujjuaq kayaks felt tippy, despite an ample beam of around 28 inches (71 cm), the strongly flared sides bulged in a graduated series of changing surface angles achieved through careful arrangement of the chine stringers, side stringers, and gunwale boards. In long kayaks, the sides often appear less carefully shaped for cross section, sometimes being flat and slablike, as in Ungava. But there were also flat-sided short kayaks; notable examples were constructed on the Belcher Islands. Their length ranged from 14 to 18 feet (4.3 to 5.5 m) with 16 feet 5 inches (5 m) common. Sheer width assured stability, with a typical beam measuring about 31 inches (78.7 cm), according to Freeman (1964:70). Bottom width was correspondingly great, at about 22 inches (56 cm). Unless loaded down, short, broad East Canadian Arctic kayaks tend to track poorly and bounce in waves (Figure 8.5).

The small kayaks have relatively short bows projecting only 2 to 3 feet (60 to 90 cm) ahead of the forefoot, whereas the larger ones have bows measuring about 4 to 6 feet (1.2 to 1.8 m) long. The short bows are more upturned and higher, around 16 inches (45.7 cm) in height when the kayak is measured sitting on a level surface. The bows are about 14 inches (35.5 cm) high and rise gradually for an easy entry. The situation at the stern is similar: more upturned and higher in the short craft. Below, the transitions from bottom to the raked bow and stern are correspondingly more abrupt in the short kayaks than in the long. Short bows maximize waterline length and the higher ends are needed since there is more pitching in waves.

The ample beam of the East Canadian Arctic kayak is notable; the design thinking behind it contrasts with that for the slender multichine craft west of Hudson Bay. The latter are made as narrow as the kayaker can handle for maximum speed, hip width being the limit. East Canadian Arctic thought is to have the kayak wide and stable while still able to attain the minimum speed required. When hunting with harpoons and darts, the minimum requirement is to be able to overtake beluga (*Delphinapterus leucas*) and narwhal, which

Figure 8.5—Ringed seal hunter Iyaittuq, 1933, northeast Hudson Bay. Photographer H.C. Gunning. © Canadian Museum of Civilization, no. 77207.

usually means a maximum width of about 25 inches (63.5 cm) at the gunwales and a bottom width of around 16 inches (40.6 cm). East Canadian kayaks can also be narrower; the minimum is around 22 inches (56 cm). When, despite their advantageous silence, harpoons and darts were given up for firearms, it was no longer necessary to be able to overtake swimming sea mammals, and minimum speed requirements decreased. Beam increased to about 28 inches (71 cm) top and 18 inches (45.7 cm) below. Atlantic Labrador kayaks generally had more rocker aft, together with a broader afterhull relative to the forward section. The Atlantic Labrador kayak is the most pronounced expression of the East Canadian Arctic design with the maximum width farthest back and the broad afterhull very shallow. This shallowness may have functioned to keep windage low in view of the good rocker aft and wide bottom.

Seals could still be harpooned from the wide hull, an event recorded by filmmaker Doug Wilkinson, who went sealing at Mittimatalik watched over by Illauq and Kalluk:

> [T]he harpoon arched through the air. With a soft thud it struck.... Instantly the seal came awake and with a terrific splash dived deep. I threw the sealskin float from the rear deck of the kayak and paddled clear. The seal reached the limit of the long line and the float bobbed rapidly over the surface of the sea (Wilkinson 1955:134–135).

When the seal stopped and surfaced, it was killed with a .22. The float was picked up and returned to the rear deck. The seal was pulled alongside and hooked by a flipper with a small gaff to hold it while the harpoon head was worked free. Then it was hauled aboard as follows:

> First I brought the seal directly alongside the cockpit, and, as I am right-handed, on my right. I held it securely by one hind flipper and with a quick shove thrust the seal down into the water and rolled the kayak over as much as I dared to the right. Keeping the kayak tilted sharply over I pulled the seal quickly up. The water removed much of the dead weight of the body and it came up like a heavy cork. As it reached the highest point out of the water, I jerked in sharply with my arm pulling the seal's body across the rear deck of the kayak where it dropped with a thud.... If it had been a big one, I would have left the harpoon head attached and towed it to shore (Wilkinson 1955:134–135).

Illauq had taught Wilkinson how to get a seal up on deck in shallow water where capsizing was no danger; he watched the retrieval through his telescope. By the 1950s, the North Baffin hull was about 28 inches (71 cm) wide. With the original 25-inch (63.5 cm) width, the action would be trickier. A second kayak alongside was of great help (Figure 8.6).

Harpooning and shooting require a little further explanation. First and most obviously, harpooning avoids loss by sinking, but early in the open water season, seals generally float, so shooting suffices. Earlier in the spring when leads open, hunters shoot from the floe edge, then retrieve seals by kayak or small *umiak*, or the

Figure 8.6—Iyaittuq steadies the kayaks while Ainalik hauls aboard a heavy seal, northeast Hudson Bay, 1933. Photographer H.C. Gunning. © Canadian Museum of Civilization, no. 77214.

wind simply blows the seal within reach of a small cast grapple, harpoon, or billhook. In winter, when there is an open current hole, seals can be shot from the downcurrent end; they will be swept in for harpooning. Around the mid-20th century in the East Canadian Arctic, shooting and retrieval harpooning became the common kayak hunting method, suspended when seals get too thin in summer and start sinking. (Later in the season they fatten up and float again.)

In still inlets at the spring breakup, a shot seal sometimes sinks through the meltwater on top and floats on the saltwater below, from whence it can be grappled up if it is beyond reach of the harpoon or a long-handled billhook. The bearded seal is inquisitive and often will approach within shotgun range for wounding in the face so that it stays surfaced since the saltwater stings in the wounds. No retrieval is necessary when bearded seals are shot while sleeping at the edge of ice pans in early summer, as a head shot kills instantly. The fast-swimming belugas and narwhals will often sink when shot, so the aim is to shoot them when their lungs are full, either just before they dive or just as they surface. The most productive method is to drive a pod into the shallows of a narrow bay by frightening the sensitive animals with shots in lieu of splashing with paddles. Even a lone kayaker can thus capture many at once. Walrus, too, can be shot in the "legs" and driven inshore to dispatch. Arduous towing is not required, as gas buoys them up later.

Finally, in 1960s Canada, the kayak was displaced by the freighter canoe with an outboard motor, which became affordable when opportunities to participate in a cash economy increased. Kayaking must always be viewed in its wider context, whether of the seasonal economic round or Arctic life as a whole. Historically, kayak hunting with firearms is a generation-long phase dating to the mid-20th century, which left a preponderance of extra-wide East Canadian Arctic specimens in museums and obscured the sleeker centuries-old designs.

From the head of Cumberland Sound and Broughton Island, a short, distinctive-looking kayak is known, probably used for retrieving seals shot from the floe edge. An example in the Canadian Museum of Civilization (IV-X-96) lacked provenance data but was identified as such a kayak in a photo taken in north Cumberland Sound (Bernier 1910:429). Very boxy, with a framework made out of milled boards nailed together, its special feature is the construction of the ends with a raked post rising from the keel and past the gunwales, which attach to it to become a sturdy round handgrip at both bow and stern. Larger canvas-covered 15-foot (4.6 m) versions have been reconstructed at Broughton Island more recently; however, as a regular kayak the form is anomalous. If only this retrieval type were known, an East Baffin variety might not be noted; however, also at the Canadian Museum of Civilization is a full-size kayak (IV-C-4550), which is just what one might expect if the little retriever had a regular-sized counterpart. This 20-foot-2-inch (6.2 m) long specimen with a 26½-inch (67.3 cm) beam has a bow constructed like that of the retriever. The cutwater strip rising from the keel has the gunwale tips joined on its sides and extends past in a rounded handgrip. This fine-looking kayak unfortunately has no provenance information; labeling it as East Baffin is an inference from form. Boas writes of "[t]he kayak of the Nugumiut, Oqomiut, and Akudnirmiut" as though only one kind existed, covering Baffinland from Clyde Inlet south. His illustrations, though evidently drawn from handicraft models, show them to be like the Hudson Strait–Atlantic Labrador variety (Boas 1888:408–409). The retrieval form of the Akudnirmiut may have been more widespread. A large version was reconstructed in the 1970s at Broughton Island as the local type, which may be due to the early demise of the full-size kayak there, perhaps due to 19th-century commercial whaling. Boas has written, "When I first visited the tribes of Davis Strait no kayak was to be found between Cape Mercy and Cape Raper, nor had there been any for several years. In the summer of 1884, however, two boats were built by these natives" (Boas 1888:408).

The historically known Polar Inuit kayaks of Northwest Greenland of the later 19th and early 20th century are flat-bottom, hard-chine derivations of North Baffin kayaks brought by migrants in the 1860s. At the time the Polar Inuit lacked kayaks. Nor did they have low tunnel-entrance heat-trap igloos, bows and arrows for caribou, or the leister, a three-pronged spear, for char, which was brought by the Baffinlanders (Rasmussen 1908:32). Regarding the introduction of watercraft, one of the Baffinlanders said:

> And we taught them [the Polar Inuit] to build kayaks, and to hunt and catch from kayaks. Before that they had only hunted on the ice, and had been obliged during the spring to catch as many seals, walruses, and narwhals as they would want for the summer, when the ice had gone. They generally went for the summer, to the islands where the eider duck hatched, or near razorbill rocks, as here at Agpat, or inland, or to a country where the Little Auks bred. They told us that their forefathers had known the use of the kayak, but that an evil disease had once ravaged their land, and carried off the

old people. The young ones did not know how to build new kayaks, and the old people's kayaks they had buried with their owners. This was how it had come about that kayak hunting had been forgotten (Rasmussen 1908:32).

Less open water, due to colder climate, and wood scarcity may have discouraged building (Gilberg 1974–1975:164, after Robert Petersen). Without the kayak for sea-mammal hunting, or bow and arrow for caribou, or leister for char, the Polar Inuit summer was a season of scarcity focused on birds and eggs. Fall returned, with sled travel and ice hunting of seals and walrus. In late winter, ice-edge hunting was done where leads opened, but starvation was common before spring came with its basking seals (Gilberg 1974–1975:160–161).

The Baffin-derived Polar Inuit kayak was small in East Canadian Arctic terms, usually about 16 to 18 feet (4.8 to 5.4 m) long by 22 to 23 inches (56 to 58 cm) wide. The flat sides had little flare. Bottom width ranged from 16 to 20 inches (41 to 50 cm). The ribs were unbent, of three straight sections. Bottom and sides met sharply. This Polar Inuit kayak is very variable, especially aft (Adney and Chapelle 1964:207, figs. 199, 200; 210, fig. 205). Signature features include a recurved Clipper bow, boxy cross section, and a broad paddle-rest piece atop the coaming. Some of the variability, such as the high stern, might result from seeing a southern West Greenland kayak in the mid-1850s before the Baffinlanders arrived (cf. Mary-Rousselière 1980:75n4). In 1853, a southern Greenland kayak was brought to Smith Sound by American explorer Elisha Kane. The kayak had been purchased in Fiskernaes (Qeqertarsuatsiaat) (Map 2) for a young hunter hired there, Hans Hendrik, also called Hans Christian (Kane 1856:23). Kane retreated south in 1855, but Hendrik stayed on, joining the Polar Inuit (Hendrik 1878:33). Passing by in 1858, Francis McClintok heard at Cape York that Hendrik was living in Whale Sound, but with neither sled dog nor kayak due to famine. The kayak skin had been eaten (McClintok 1860:122).

The influence of West Greenland kayak design began about 1910 via a trading post and took over within a couple of decades. Some Baffin elements remained, however, like a flatter bottom (cf. Holtved 1967:77, fig. 55). In adopting West Greenland characteristics, the Polar Inuit actually returned to their old kayak roots, for in 1921 kayak and gear remains were found far north in Washington Land in Morris Bay, latitude 80° 10', by Lauge Koch. Upon study at Disko by four knowledgeable Greenlanders, these objects were dated to the 16th or 17th century by Therkel Mathiassen (1928:192–212). The kayak, apparently cached, was estimated to have been at least 15 feet 7 inches (4.75 m) long. Width was estimated to have been about 19.5 inches, given an 18½ inch (47 cm) *masik*, or arched cockpit front beam, and gunwale sections notched, rather than mortised through for the crosspieces. The whale rib *masik* has a finger-grip depression cut in underneath to help the kayaker pull himself in (Mathiassen 1928:192, fig. 3), a useful feature still included in some Greenland kayaks.

The Baffinlanders did not bring the *umiak*, already rare at home, to the Polar Inuit. Migrations took place in the spring by sled. The Polar Inuit have ended up being the only Inuit who regularly hunt by kayak in the 21st century; they focus upon narwhal (cf. Silis 1984).

Lucien Turner's observation from Ungava of the kayak's key role in a man's life may help to contextualize kayak use. In the following, sealskins are the critical material required to construct the basic equipment for Arctic survival: kayak, *umiak*, and tent.

> A young man starts out in life with a gun and ammunition with which to procure game. If he has the energy to become a successful hunter he will soon be able to make a kaiak, and thus procure the marine mammals whose skins will afford a covering for an *umiak* and in the course of time additional skins for a tent. These possessions usually come in the order laid down, and when they are all procured he is generally able to have others under his direction assist in transporting them from place to place; and thus he becomes the head of a *gens* or family, including his brothers and sisters with their husbands, wives and children.
>
> Some of the men are too improvident to prepare these skins when they have the opportunity, and thus they are unable to own a kaiak, which prevents them from providing themselves with the *umiak* and tent. These persons must live with others or dwell by themselves and pass a miserable existence, scarcely noticed by their fellows even during a season of abundance (Turner 1894:240).

KAYAK MAKING

Gathering wood for the framework. Materials for kayak construction were often gathered well ahead of time and from a distance. Driftwood was commonly used for the framework and was gathered in the spring by dogsled. Journeys were also made south to the tree line for live timber around Kuukjuaraapik or

Great Whale River, the head of Ungava Bay, and southern Atlantic Labrador. In a fascinating procedure, northwest Labrador Peninsula Inuit sometimes made sleds at the trees, dog-teamed back north, and then converted the sleds into kayaks. The reverse procedure also occurred. Trade lumber had been used to construct East Canadian Arctic kayaks since the Hudson's Bay Company ships began to supply Hudson Strait in the late 17th century. Trade lumber then became uncommon until the 19th and 20th centuries, when whites returned en masse to the Arctic to pursue commercial whaling and Christian missionizing and to establish northern fur-trade posts. Old Piugaattuk of Igloolik said that white whalers were the main source of wood for the Foxe Basin–North Baffin region from Igarjuaq near Mittimatalik in the north and from Roe's Welcome Sound for the south (Piugaattuk 1991).

Tools. Traditional Inuit tools, such as the elbow adze, were improved with iron blades. Inuit crooked knives, gravers, and chisels all improved with the trade in metal. The European brace and bit, as well as chisel and hammer, were significant tool acquisitions, although their Native counterparts retained advantages when metal-bladed. The bow drill remained in use for small holes, enabling one-handed operation. The graver was used for finer splitting jobs, such as in bone and ivory work. Steel needles were superior to bone ones; the triangular glover's needle was best for the heavy skin sewing of the cover. For cover skin preparation, the *ulu*, or half-moon-shaped woman's knife, remained the primary tool, though its blade changed from stone to steel, usually originating as a circular saw blade. Skin-dressing scrapers were also given metal blades. The man's traditional flensing and butchering knife had its triangular blade made of saw steel, too. In the recent tool kit, the trade pocketknife is a standard item. Last, but not least, is the file, not just for sharpening but also for shaping wood, ivory, bone, antler, and metal.

Knowledge of different woods was well developed, and wood properties were matched to the requirements of different parts of the kayak framework and gear (cf. Petersen 1986:18–19). Those living far from forest stands had minimal choice in wood type; however, the spruce, commonly recovered as driftwood, is good for most kayak construction purposes. Driftwood works well when seasoned in saltwater due to a leaching effect, which leaves the wood easy to cut, less brittle, and less prone to cracking.

Top Framework Parts. The two *apummaak*, or gunwale planks, are the principal members of the East Arctic kayak framework. If wood of sufficient length is available, the *apummaq* (singular) can be constructed from one piece. How a single large driftwood log was hewn and split into two gunwale planks has been described in detail for a Belcher Islands two-seater reconstructed for the Canadian Museum of Civilization (Guemple 1966:164–167). However, such wood was usually unavailable for gunwales longer than the usual trade lumber measuring 16 feet (4.9 m) in length, making it necessary to join two or three shorter lengths together. Imported lumber was in common use in many places by the late 19th century.

A hook scarf was traditionally used until iron nails replaced the lashing and pegging to hold the joint. Then the little stabilizing hook or jog in the interface became unnecessary. In joining the gunwale sections, craftsmen take advantage at the scarfs to angle the sections to achieve the desired reverse sheer to the gunwales. Because the gunwale planks are leaned out and the sheer decreases when they are spread when shaping them on the flat, extra curvature upward is included. This marked "humping up" amidships, coupled with the wide gunwale planks, means that to make them in one piece requires wide boards.

When the leaned-out gunwale planks are spread, the bow and stern rise higher, so allowance is made for this change. Bow and stern are shaped to curve upward, the latter for only about 8 inches (20 cm) of length. Before the short stern turn up, often only slight, some gunwales curve down to decrease height and windage aft to counter weather cocking with the flat-bottom design. The bow is shaped to rise smoothly for a greater distance, about a fifth or sixth of the length overall. In longer kayaks, this rise is gradual, sometimes leveling off at the tip. In shorter craft (16 to 18 feet long), the bow is more upturned to surmount the waves and catch more wind, which, while generally a disadvantage, can be an asset when running downwind by counteracting the stern's tendency to be blown around. These design features result in a rather complex shape to the gunwales in profile. The gunwale planks are about an inch (2.5 cm) thick, or slightly less with the recent standard lumber from the south, and about 4 to 5 inches (10 to 14 cm) wide in the deepest middle section, tapering to the ends with the bow tip often just 1 to 1½ inches (2.5 to 4 cm) thick. Toward the stern the planks are kept deeper. The kayak ends are handholds, so a variant stern form has a slight top bump for a grip stop. The stem is made just slim enough to fit the hand, with the exception of the deep North Baffin form. When the gunwales are made from a log, the large end can be used for their sterns, providing enough wood to allow the gunwales to curve in and turn up to form the stern post halves (Guemple 1966). Bow construction from a log can be similar, though less pronounced, in the turnout of the sides from the end. The stern can be built around an end block, which is shaped to become a handgrip post slanting up at the back; the gunwale ends are left simply straight and mortised into the block sides. In North Baffin, a postless block is used for a plain stern without any handhold.

Stern construction is more variable than that of the bow; however, there are a few common solutions and since they tend to be regionalized, the origin of provenance-poor museum specimens can often be identified based on their particular stern form.

The two gunwale pieces are matched in shape and flexibility to bend around the crosspieces in the same way for hull symmetry. Sections are thinned down as necessary. When the gunwales are made of boards, instead of being sculpted from raw timber at the stern, they have to bend especially strongly in the aft quarters for a full outline. In shorter kayaks, which need more curvature in this section, the gunwales might be thinned down. Numerous close-set cuts across the inside surface are another means of increasing flexibility.

The gunwale boards are marked for deck beam and rib locations. For beam location, small holes are drilled through planks while they are matched together. The moderately arched *masik*, by far the heaviest crosspiece, is located first amidships, or a little aft if preferred. The straight *itsivik*, or cockpit back beam, the next heaviest, is placed to give just enough space for the knees to clear the *masik*, legs bent. This more relaxed entry and exit make for a longer cockpit than straight-legged sliding under the *masik*. The *itsivik* is commonly about 23 inches (58 cm) away from the *masik*. A lighter and less arched *masirusiq* beam or two (*masirusigiik*) rise to support the deck stringer as it rises to the top of the *masik*. They are typically made from curved branches. The *masik* is best when made from a heavy curved limb. The rest of the deck beams fore and aft, the straight *ayaat*, are spaced regularly about a handspan apart, or a bit more in kayaks dating after the early part of the 20th century. Ribs, or *tikpiit*, are spaced like the top crosspieces, or a little closer since they are lighter. At Ivuyivik, at the northwest tip of the Labrador Peninsula, I was told in 1960 that rib spacing was formerly narrower, measuring about 6 inches (15 cm, or the thumb-to-forefinger spread), but had recently increased. In all such details, individual, regional, and historical variations exist. On the lines drawings for many of the East Canadian Arctic examples, the size and spacing of parts are indicated on the structural cross section.

The rib locations for drilling mortise holes are marked along the gunwale plank bottoms. In order to avoid weakening the planks,

Figure 8.7—Ainalik (right), assisted by Iyaittuq, installs a new *niutaq*, or cutwater strip, which will be bent down to meet the *kuuyaq*, or keel, in a scarf join. Photographer E. Arima, 1960. © Canadian Museum of Civilization, no. J16409.

Figure 8.8—Rebuilt framework of Ainalik's kayak raised above the reach of dogs, who would eat the sealskin lashings. The *paaq*, or cockpit hoop, will be added later. Photographer E. Arima, 1960. © Canadian Museum of Civilization, no. J16403.

these locations usually do not coincide with those of the deck beams. The upper mortises for the crosspieces are rather shallow, V-shaped notches cut a little below the gunwale tops so that the deck stringer over the crosspieces will be flush or only slightly higher than the gunwales. The rib holes are made as large as plank thickness permits while keeping enough sidewall strength—half an inch (1.3 cm) or a little greater in diameter, and maybe a good inch (2.5 cm) deep. They are bored with brace and bit when the planks are matched together, or later when the ribs are being inserted.

The planks are temporarily nailed together for matching or, alternatively, are held by wooden dowels that can be planed over. The top inside corners are beveled off to keep them from sticking up into the cover at the sides of the deck. It is possible to do this chamfering, and also the crosspiece mortise cutting, while the planks are still together as long as they are matched inside out. The heavy cockpit front beam might be as much as 3 to 4 inches (7.6 to 10 cm) wide by a full inch (2.5 cm) or more thick. The up curve, measuring 2 to 3 inches (5 to 7.6 cm) was considerably less in the 19th century, while manhole length was shorter, suggesting that the legs were kept straight to enter and exit, as in Greenland. In the past, the cockpit hoop was not tilted up as high—perhaps so that the waterproof gutskin jacket could be tied around it to keep the waves out. The less curved crosspieces in front of the *masik* are made of a branch around 1¼ inches (3 cm) thick, trimmed partially away above or below. The heavy, straight cockpit back beam is about 1 by 3 inches (2.5 by 7.6 cm), laid on the flat and preferably a little higher than the rest of the beams behind so that the aft deck rises slightly to encourage waves to wash off. Its leading edge is usually shallowly scooped for the kayaker's back. The other ordinary crosspieces are about ¾ by 1½ inches (1.9 by 3.8 cm) with corners rounded off. They can be smoothly ovoid or rectangular depending on workmanship.

The anthropometric sizing of kayaks is approached differently in Canada and Greenland. Although both kayak and kayaker are always considered, the hull is of primary importance in Canada, while in Greenland, the kayaker's body measurements are addressed first. In Canada, kayak length is reckoned by the number of skins needed to cover the hull, a nonanthropometric measurement with economic import. In Greenland, length is reckoned by the kayaker's armspan. Width in Canada is figured anthropometrically by forearm, handspan, and finger thickness measurements. In Greenland, width is an important consideration in rolling, at least today on the Central West Coast. Hull depth amidships in Canada is indicated anthropometrically by the vertical thumb-to-index-finger span inside the framework at the back of the cockpit, which provides enough room for items loaded in the back, such as a small seal. Total hull depth is a compromise between having little freeboard to reduce windage and more freeboard to achieve higher reserve stability. In Greenland, depth is significantly shallower than in Canada for less windage; in West Greenland today, depth is as shallow as possible in order to facilitate rolling and bracing techniques.

While dimensions are often reckoned in terms of body parts, and thus might be labeled anthropometric, measurements may or may not be tailored to the individual kayaker. Performance characteristics are the actual determinants of kayak dimensions. The units of measurement used, whether body-based or otherwise, must be distinguished from the way in which they are employed to achieve the desired design.

Top Framework Assembly. Assembly of the framework begins by joining the stem ends of the gunwales, which are given flush-fitting faces inside, though not on top, where a groove is left for sewing the cover seam later. This stem join can be lashed with turns of sealskin line through paired holes connected by a counter-sinking groove outside, or might be nailed or screwed together. If a screw or two are used, brass is recommended for corrosion resistance at this critical joint. Any metal nails used in assembly have to be clinched or they will work loose. After the stem is fastened, the gunwales are spread around the cockpit front and back beams and drawn together at the stern using a line loop pulley. If fitted, the stern block is inserted, and the gunwale ends bedded in its sides. The crosspieces have their ends sharpened to fit into their groovelike mortises, which are cut in the inside of the gunwale planks. The latter are flared as the builder sees fit, usually less than the planned overall side flare, which typically ranges between 20 and 30 degrees from vertical. The rest of the crosspieces are then set in, usually from the ends, sizing as necessary to get the desired top outline. Crosspieces are fastened by continuous lashings, *qilarutit*, preferably of the strong skin of the bearded seal, beginning at the cockpit beams and going to either end. At each beam, the line commonly comes along below on the inside from the previous one, goes out through the gunwale plank by a hole below the beam, comes back in through a hole just above, goes down through the beam by a hole a little inboard, and loops around itself where it first went out through the gunwale before continuing to the next beam. On the outside, the line is countersunk in a groove connecting the pair of holes in the gunwale plank.

In a variation, the upper hole is bored from the top inside chamfer on the plank to emerge outside at the lower hole so that no countersink groove is needed. Crosspiece lashing can also be by separate ties at each juncture. Another occasional practice is to have a few ties linking the gunwales with the line end wound around the looped strands to pull in the planks as Spanish windlass clamps, *nutsurutit*. Such ties can be made in the lower part of the gunwale boards under a crosspiece, with the line end wound around both the tie and the beam for a strong flare, maintaining pull. This mechanism, widespread in Canadian Arctic construction, is seen also in older Greenland kayaks (Petersen 1986:60). Now Greenland deck beams are tenoned through the gunwales tightly with wedging. Tenoned cross members seem to proliferate with iron tools; a similar development is occurring in north Alaskan *umiak* construction.

The deck stringer, or *tuniqjuk*, fore and aft sections are fastened with continuous lashing; their ends at the cockpit beams are housed in top corner slots, while the bow and stern ends are left free or fitted between the gunwales, or to a stern block, if present. The lashing line comes underneath from the previous juncture, under the beam and up one side of the stringer, then goes down through a hole in the stringer and out below, back past the beam, bending around under the line section coming from the previous tie to draw it up tight by next passing over the beam on the same side of the stringer as before, and finally going on to the next beam, but cutting diagonally across under the stringer to its other side. On the stringer surface, a groove is cut from the side at an angle to the lashing hole to countersink the line. These slots on the stringer alternate from side to side at each beam. The same continuous lashing method is used on the keel and the chine stringers. The lines can be seen zigzagging along them inside. The procedure is simpler than it sounds here. East Canadian Arctic framework assembly is a comparatively easy procedure, although getting the best shape is another matter. As elsewhere, good craftsmen are few and highly sought after for both construction and consultation.

Bottom Framework Construction. The framework is turned upside down to construct the bottom. The ribs are the first to be added. Measuring about ½ by 1 inch (1.3 by 2.5 cm), they are rectangular in section, with just the corners dropped or smoothed into an oval shape. As in the case of the deck beams, the ribs were rounded in the past. Their raw lengths include ample spare for trimming to size during the fitting to the framework, with the rib bending in the process of shaping the hull bottom. The ribs bend quickly between the flat bottom section and the leaned-out sides, with perhaps a notch or two cut on the inside to facilitate bending. When the wood is pliant, notching can be dispensed with. The ribs were heated in hot water, sometimes with oil added, and bitten to weaken the wood fibers. Typically, at first just one side of the rib is bent and stuck in place on the upside-down spread gunwale boards. Occasionally a rib cracks at the bend. It might still be salvaged by wrapping with sinew, often incorporating a small reinforcement across the break. But a break right through requires a fresh start.

An alternative three-piece construction is known with separate bottom and side sections nailed together. Such a construction may occur when replacement is needed, but bending was not possible due to poor quality wood, or when heating was too troublesome to arrange. A shoulder is usually left on one of the ends at the joins; as with other details, common-sense practicality rules. Rib insertion usually starts at the cockpit back, where depth might be gauged by the thumb-to-little-finger spread between the top beam and the rib. The bending on just one side at first enables the depth to be easily set by cutting the inserted section to length. Succeeding ribs are similarly set along the same gunwale, eyeballing for the bottom profile; a straight edge or taut line may be used. Then the location of the bends of the other side can be readily estimated to give the desired bottom shape. The ribs are removed, reheated, and given their second bends. Reinsertion of the fitted sides and transverse leveling of the flat bottom sections allow easy determination of where to cut the other sides; uniform depth in the drilled rib mortises is helpful. Finer trimming is done as desired, but a little leeway exists in rib fitting since a little shimming can be done if necessary when the stringers are put on. Since the forefoot is at the beginning of the bottom and the deepest section of the hull, the nearly V-shaped rib there with just a brief bottom section might be inserted early to guide the bottom line formation. As the other depth-setting rib at the cockpit back is shallower, the line slants gradually closer to the gunwale boards, with the hull becoming quite shallow aft from a combination of the gunwales curving down with reverse sheer and the bottom curving up in a gentle rocker from amidships. The bottom may curve up more quickly at the stern in order to join the gunwales for an angled end profile; alternatively, the bottom can meet the gunwales by continuing the gradual rocker, a smoother form seen more frequently in long kayaks.

The rib-bending process affords control over the bottom shaping by handling the depth and width dimensions separately in stages. To hold them in place, the rib ends are often pinned in their mortises with small wood or metal nails (the latter more common by the mid-20th century), countersunk outside, so as not to contact the skin, and clinched inside. Alternatively, transverse sealskin ties between the longitudinal framework members at several places along the hull can be used for rib retention. To finish each tie, the line end is wound around its previous loop for the tightening effect. All of these sealskin lashings help to make the framework more rigid, while at the same time allowing it to give slightly, rather than break under stress; moreover, should the framework break, the multiple ties help hold the kayak together.

Before whalers wiped out the bowhead whale in the late 19th century, long bendable baleen was used to construct some of the smaller kayak parts, such as the hoop, ribs, and stringers. Baleen ties occur in older frameworks. A deck-gear retainer baleen strip bent up at the sides was found inside a South Baffin kayak, now at the Royal Ontario Museum. It was collected by James W. Tyrrell at Big Island, where he manned one of the observation stations along Hudson Strait in the late 19th century. This kayak has a sealskin line tied across each bottom corner; the same bend-holding method might be expected with the ribs. A hole near the top of one upright arm and three or so notches on the other indicate a tie across the top adjustable for height, possibly to retain a bearded seal float, since the deck equipment strip was several feet behind the cockpit, based on the vestiges of its sew-down points. The corner bend ties form openings to hold the lance, bill hook, or some other long-handled implement. The same baleen deck-gear holder was seen in the 1820s at Igloolik (Parry 1824:facing 274 and 508), and may be the source of the South Baffin example.

For deck equipment forward, Tyrrell's kayak has ivory eyes in the sides near the gunwales, which may have anchored a deck line with upstanding toggles, as in the West Hudson Bay craft. Another example is the North Baffin kayak in the Royal Scottish Museum in Edinburgh. That specimen also has many baleen ribs, stringer sections, deck beams, and cockpit hoop, and is the closest thing to the mythical whalebone kayak known. Baleen, or *suqqaq*, being more water- and rot-resistant than sealskin or sinew, was favored for ties, though it did not come in the lengths needed for the continuous lashings for deck beams, keel, and stringers. Whale skin was not used for covering watercraft, apparently because it is too permeable.

In the prow, several V-shaped ribs support the cutwater strip, or *niutaaq*, which in East Canadian Arctic kayaks is fairly thick and wide, measuring about 1¼ by 2¼ inches (3.1 by 5.6 cm). It has depressions cut in the inner surface to house the apex of each V rib, which may have two close-set kerfs to make the acute bend. Paired struts may be used instead, tied together with the *niutaaq*. The other ends of either V rib or paired struts insert in the gunwales. The *niutaaq*'s outer side is rounded, often with a little sharpening in the waterline area. Although this cutwater proper may still look a bit blunt, the pronounced rake makes it fine enough in the flow. The *niutaaq* is bent, hence its name, around the forefoot and scarfed onto the keel, or *kuyaaq*, which is often thinner to match the chine-stringer thickness in order to give a flat bottom.

The cutwater and keel can also be in one piece, particularly when the kayak is not too long. The cutwater strip has Spanish windlass ties to the gunwale planks between the V ribs or struts. Fastening beyond the forefoot is accomplished through the continuous lashing of the *kuyaaq* section to the ribs. The ends of the *niutaaq-kuyaaq* can lie against the bottoms of the gunwale planks at stem and stern right to the tip or stop short, housed flush in slots and fastened securely by ties and/or pegs, nails, or screws. The chine stringers, or *sianiik*, are usually wider than the keel strip, measuring about 1 by 3 inches (2.5 by 7.6 cm). They often have long diagonal darts cut from the inside almost to the outside and nailed to close like scarf joints. Their ends are fastened onto the sides of the keel and faired in. In the Hudson Strait region, the chine stringers are shallowly grooved across the top

Figure 8.9—Several South Hudson Strait kayaks at Kangiqsukjuaq, a rich hunting locality, 1897. Photographer A. P. Low. Courtesy National Archives of Canada, no. PA-051468.

to have an outside lip about ¼ inch (6 mm) high to retain the ribs. Where this grooving and lip are lacking, as on East Hudson Bay south of Qikiqtauyaq or Cape Smith, a full-length side stringer is added to retain the ribs at the sides. With lipped-chine stringers, only a partial light side stringer is nailed on in the deeper forward hull from about the forefoot to the cockpit or just past it to keep the skin cover off the ribs. Farther aft, the hull becomes shallow enough that the space between the chine stringer and the gunwale board can be bridged by the skin without risk of caving in onto the ribs. Extra floor stringers might be added between the keel and the chine stringers, sometimes making the bottom quite solid when all the lathes are wide. These side and floor stringers end freely. Transverse floor boards are usually added between the ribs to keep the kayaker out of contact with the wet skin and possible bilgewater. Resting on the keel and chine stringers, these thin boards extend forward under the legs. All outside surfaces that have contact with the cover are planed smooth.

Skinning. The kayak's cover, or *amiq*, meaning "skin," is primarily constructed by women (Figure 8.10). The kind of sealskin used is a major consideration, though a very limited use of caribou occurred inland on the Ungava Peninsula between Ungava Bay and northeast Hudson's Bay, where a small design is known via reconstruction only (CMC IV-B-1445). Bearded sealskin, thick and strong, is preferred for most East Canadian Arctic kayaks. Ringed sealskin is widely used, as in North Baffin, when bearded seal is lacking or would break the light bottom framing when it shrinks. Ringed seal may also be used if lightness is desired for frequent portaging inland or across necks of land, as on the Belcher Islands. Caribou is a lighter skin, but is weaker and less water-resistant.

Since ringed sealskins are the smallest sealskins used to cover East Canadian Arctic kayaks, their preparation entails extensive thread-making and sewing. For the same size kayak, about twice as many ringed sealskins are needed as bearded sealskins. Four large bearded sealskins will cover a kayak measuring 21 feet (6.5 m) long. That number of bearded sealskins may have been one reason for the 19th-century popularity of the 21-foot kayak. Longer kayaks are common from the later 19th century in the Hudson Strait region. Voyaging to trading posts also became common at that time, fostering the construction of large kayaks measuring 26 feet (8 m) long.

The preferred kayak covering used in Greenland is the skin of the harp seal, also known as the saddleback or Greenland seal; this type of skin is also occasionally used in Canada. H. C. Petersen

(1986:29–38) discusses the types of cover skins used in different parts of Greenland as well as the whole covering process. Petersen's description is applicable to East Canadian Arctic kayaks, except for some minor details. The hooded or bladdernose seal (*Cystophora cristata*) mentioned for South Greenland, where harp seal is difficult to find, only occurs in the Canadian Arctic on the Atlantic coast. The skin of the hooded seal is more permeable and thus less waterproof than that of other species. Whale skin is also permeable and not sufficiently durable, according to Petersen, who cites an experimental case in which a kayak was covered with whale skin (1986:30). This permeability and lack of durability explain why this sea mammal was never used for kayak construction. Other whales, such as the killer, bowhead, humpback, and sperm whales, presumably pose similar problems. Walrus hide is too thick and heavy to use unless it is split, a preparation technique not practiced in the East Canadian Arctic. In parts of Alaska and northern Asia, female or young walrus skin, which is lighter, smoother, and stronger than other kinds of walrus skin, is used for *umiak* covers, but only to a limited degree for kayak covers.

Figure 8.10—Women sewing the deck seam of the *amiq*, or cover. Four and a half bearded sealskins will be used to cover this kayak. Left to right: Ilisapi, Litia, Kuara, Ilisapi, Maata, Piatsi (?). Photographer E. Arima, 1960. © Canadian Museum of Civilization, no. J16411.

Sealskins have potential leakage points at the foreflippers, mammaries, navel, genital opening, and weapon holes, which have to be sewn shut or patched before use. To reduce the number of holes, a skin can be trimmed at an angle to position a fore flipper at the edge. The hind flippers and tail are trimmed away, often at a slant to match the oblique angle of the next skin. As a rule, skins are positioned nose end forward. When the hair is cut, short remnants remain; they should lie with the flow of the water. In North Baffin a more complicated assembly zigzags the skins (Mary-Rousselière 1991:49, fig. 3).

Another method of skinning the bearded seal is to make a spiral cut from a mouth corner back and over the neck to the shoulder and across to get its foreflipper to the edge, then to cut diagonally down the front to the other side to give a longer piece than with the straight ventral cut. This method is handy for wrapping the tapering end sections.

To prepare the cover skins, the fat and flesh are cut and scraped away from the inner surface. Skins are rinsed in seawater and soaked if they become dry at any stage. To dehair the outer side, the skins are usually ripened to loosen the hair by keeping them in a sealskin poke—now the ubiquitous 45-gallon fuel drum—for about 10 days. Hot water is also used in East Baffin (Boas 1888:520). Besides removal by plucking or scraping, the hair can be cut off close to the skin; this way, the epidermis is retained for a more waterproof black or brown cover. Scraping off the dark epidermis produces a creamy white kayak, seemingly conspicuous but actually camouflaged like floating ice. Aesthetics are involved, too—the black and white extremes are preferred to the intermediate dull browns, although species and season are factors that are routinely considered.

The famous waterproof double seam is used to assemble the skins into a single piece as long as the kayak. The skin edges overlap by about a finger width. There is some variety in the way the seam is arranged, but a simple single lap is most common. The seams are sewn with the skins suspended from a raised cross beam; the lifting lines are looped around skin gathered over a sharp-ended toggle inside. Caribou back or leg sinew is laboriously braided into three-strand thread, although bowhead whale sinew has been used in the past. White whale sinew, though shorter, might substitute for caribou, but seal sinew is too short. More recently, in areas where caribou sinew is not abundant, cotton is sometimes used for one of the strands. Sinew thread is strong and swells when wet to stop the sewing holes. The stitches are embedded in the thickness of the skin using a running stitch with a locking loop back. The thread is inserted into an existing needle hole where possible to reduce piercing of the skin. Sharpened glover's needles that are triangular in section are used. The top side of the skin edge is sewn down to eliminate the ledge that would be formed by the thickness of the skin if the stitches were put in the underside. On the outside surface, the edges of the transverse seams also face aft. A final detail to minimize drag is to have the outside of the double seam flat with the overlap bunched up slightly to the inside. The cover is stretched tightly over the framework, lengthwise first with the front end sewn into a little pocket to fit over the pointed tip of the bow and a line attached at the rear to facilitate pulling. For stretching transversely, a strong braided sinew line is used, zigzagging down the deck with embedded stitches taken in the inside of the cover on alternate sides to achieve an effect similar to shoe-lacing. Men do the cover stretching, which requires strength. Then the women sew the cover together on top with patch pieces added to fill gaps. Sometimes in this deck sewing an overcast stitch is used for the second seam, rather than the more painstaking embedded and locked running stitch.

An Igloolik woman's view of kayak covering is provided by Martha Nasook. The sealskins used are taken before the molt; otherwise water seeps through. Dehairing is normally by scraping and/or plucking after aging to loosen the roots.

> [I]n emergency cases you may shave the fur off but they [sealskins] are smoother and don't make any noise in the water when they are aged; the skin is all smooth and flat. After taking the fur off you take the first layer of the skin. Then you have to rinse them, and you are ready to cut them up. You just don't cut the skins in any way. You have to cut it from one side of the head then onto the other side of the skin towards the lower part. You cut it as straight as possible by having someone holding on to it or you can tie it onto something. The knife has to be very sharp because you have to cut it with one stroke so you have no curves on the edges. After all the skins have been cut, starting from the *usuuyaq* [bow] of the kayak you start sewing them together starting from the smaller skins and the larger skins are sewn around the middle part of the kayak. If the skins were not cut properly additional skins were required to be added on to the top. Other times if they are cut properly you didn't need to add any skins to the top. When everything is ready to be sewn, you have to chew on the edges just like you would in sewing the *kamik* soles. You chew out all the water from

the edges; this is so that the stitches will become waterproof. Some young people would do the chewing around the edges. Women would have braided sinew together of caribou leg sinew. Also they would make finger protectors [sealskin tubes] to fit around all the fingers…when you are using the braided sinew it is very hard on one's hands. Then after preparing everything you hang the skins, and they start to sew them together. The stitches had to be very tight to make them waterproof. Some stitches would be called *sukkaittuliyuq* when they can't make the stitches tight enough. The stitches would be sewn in the same manner as you would sew waterproof *kamiks*, doubled back on each stitch. The skin edges would be slightly overlapped and sewn from inside and outside…. The front of the kayak, *usuuyaq*, is padded over the frame before you put on the cover, sometimes with caribou skin, so that even if it hits ice the cover will not be easily punctured. Once the sealskins are sewn together, the man would stretch the cover onto the frame and tighten it as much as possible by using lines. Then you would start to sew the top….

Kayaks lasted awhile as long as they didn't get eaten and were looked after well. When the cover was dry seal fat would be applied to the seams, and if it were to be punctured a new piece would be added on making sure to waterproof it well. After the introduction of regular sewing needles those curved needles were very handy to use for mending the holes of kayak. Seal fat often had to be applied on the seams and surface of the *qayaq* just like they did to the waterproof *kamiks*. The water does not soak through the skin as easily when the fat has been applied. It becomes waxy when it dries up; it acted as a paint (Martha Nasook 1991, IE-175, interviewer and translator Leah Otak).

Aged seal oil turns into a varnish, hence the paint reference. Felix Alaralak, another kayak informant recorded at Igloolik, also mentions the significance of noise as a factor in kayak construction.

The skins on the *qayaq* could see two seasons but some would last only for one. When kayaks were being covered I found it depressing and melancholic as the work got to be done at night. I always wondered why they had to be done at night when the temperature was getting cold, the sun no longer shone, and I was getting sleepy. The cover on the kayak had to be completed once they started with it. In the spring when the seals were basking on the ice, the skins were collected to be used for the cover. When the time came

to cover the frame, the skins would have slightly turned, *ujjaq*. Of course, when there were not enough skins, they would have to make do with skin recently caught. It was preferred to use the *ujjaq* as the fur could be peeled right off; otherwise with fresh skins they would have to shave the fur off. This kind of skin, they say, makes too much drag when one is trying to approach an animal, *quluraayattuq*; it tends to make noise as the kayak skims along the water surface (Felix Alaralak 1990, interviewer Eugene Amarualik, translator Louis Tapardjuk).

In contrast to Alaralak's melancholy, Uyarasuk, a major source of information on kayaks at Igloolik, says the women were joyful while sewing:

When the frame of the kayak is completed so that all it needs is the skin, the frame is coated with oil from blubber so that it is slippery when the skin is going to be stretched on taut. The area at *usuuyaq* [bow] would already have been sewn so that all you would have to do is slip it in. In this manner the process of skinning a kayak begins; done this way the skin when sewn together will be tight. The women who were sewing the skin onto the frame used to be so jubilant as they would know then that the days were coming when they would be able to get fed with fresh meat once again (Uyarasuk 1991, IE-179, interviewer Maurice Arnattiaq, translator Tapardjuk).

To complete the covering, the last part of the framework, the cockpit coaming, or *paaq*, is added on outside. It is egg-shaped, flattened toward the larger back end. Size and shape can be simulated by extending the arms in a curve with the fingers meeting at the front. Hoops used to be smaller, but by the mid-20th century the opening was typically around 21 by 25 inches (53.3 by 63.5 cm). By then, the D-shaped coaming with a separate straight back section became common. It was easier to make, but not a good shape for sealing in the cockpit with the waterproof gutskin jacket. The advent of hunting with guns instead of harpoons and darts seems to have ended cockpit sealing. The acquisition of motorized whaleboats and Peterheads, which superseded the *umiak* from about World War I times on, was also a factor in ending cockpit sealing by lessening the importance of kayak hunting. The straight-backed cockpit coaming afforded corners on which gun barrels could rest. The East Canadian Arctic coaming is deep compared to that of other kayaks, with about 3 inches (7.6 cm) typical. There used to be a slight lip around the top to retain the jacket edge. The cover lashing line is countersunk in grooves linking paired holes, which are close-set outside and diverge to the inside. As with general lashing practice, each pair of holes is slightly angled across the grain.

Deck cross lines and loops of bearded sealskin line finish the kayak. There are two lines before the cockpit and one, sometimes two, behind. The lines can be made from a single length of skin running back and forth between the gunwales. Single lines can also be installed toward the stern and the bow several feet from the lines at the cockpit. Small loops hold weapon and implement tips more exactly in position; their other ends are tucked under the cockpit lines. A common setup has one loop on the foredeck just right of the midline for the heavy *igimaq*, or *qaatilik*, harpoon and a pair of loops aft for the killing lance, or *anguvigaq*, and the long-handled billhook, or *niksik*. These loops are about 3 inches (7.6 cm) across and arched. They are formed of a line folded back a couple times to make several strands. Then the end of the line is wound around the strands to wrap them. The ends of the loops are sewn down to the deck. The loops are angled outward to facilitate insertion and withdrawal. Attached at the side of the cockpit on the right or left, depending on hand preference, is a harpoon rest, a long toggle triangular in profile that is level on top when lying on the sloped deck. A protuberance on the deeper outer end retains the curved ivory foreshaft of the heavy harpoon, which lies ready for use by the cockpit, pointing backward and upside down. The mainshaft rests on either a holder built onto the line rack side or a separate notched support, both secured by the cockpit front deck lines. In a sense, a kayak is not truly done until it is fully equipped.

After being covered, the new kayak has to dry a couple days or more, usually raised up on a high stand safe from dogs, who are unable to resist its delicious aroma. Then the kayak is well coated with aged seal oil for a varnish. The seams are also coated and may be stopped up with a thick grease as well. Up to two or three days' use in the water is possible before waterlogging and seam failure. Usually a kayak doesn't stay immersed that long. Cover life varies with use, but with care and periodic oiling, a cover can be used for two or three seasons. Re-covering annually is ideal. Since the bow and stern tend to wear first, at times they alone are re-covered, which can create ends that are darker than the main body of the kayak. The effect could be accentuated by putting black skin on the ends of a white-skinned hull to create a dramatic contrast. Cover life usually ends because of failure at the seams as the thread cuts into the skin.

The kayak may be stored on the kayak stand, the *qayaqvik* made of paired pillars of rocks, or a scaffold if wood is available. Nowadays, storage atop the house is handy. Kayaks must be elevated to prevent damage by dogs and other animals, as well as to keep them from being completely covered by snow. Normally stored upside down, the kayak has to be well tied to the stand and the whole weighted down against the wind. Burying for winter and other long-term storage was done on occasion, but was not common.

SPECULATIONS ON DESIGN ANCESTRY

The earliest probable kayak evidence found to date is a slender paddle blade and what appears to be a rib of a flat-bottomed kayak. These materials are from Qeqertasussuk, southwest Disko Bay, West Greenland (Grønnow 1988:37, fig. 15). Qeqertasussuk is a Saqqaq culture site that dates from 2400 to 1200 B.C. and is notable for its excellent preservation of organic materials in permafrost. The probable rib specimen was recovered in six pieces. It is U shaped, with a flat bottom; the sides bend up gently—one arm is upright and measures about 8.7 inches (22 cm) long while the other is angled slightly outward. Both are broken. The specimen is 13.8 inches (35 cm) wide. Presuming that the artifact is a rib, the shape need not mean the hull bottom was flat, since the keel stringer can be deeper than the flanking stringers. Nevertheless, with a U-shaped rib, the usual intent is a more or less flat-bottomed design.

Far to the west on the Asian side of Bering Strait is the earliest evidence of a complete kayak shape. The Okvik–Old Bering Sea (300 B.C.–A.D. 500) site of Ekven Cemetery on the Chukotka coast yielded an ivory kayak model (cf. Arima 1985:23–25; Arutyunov et al. 1964:342; Zimmerly 2000:3). With a shallow arch bottom rounding up into well-flared sides, the design is essentially flat-bottomed. Most noticeable are the odd, horizontally forked bow and stern. There is a two-millennia gap between the Okvik–Old Bering Sea model and the older Saqqaq rib; nevertheless, these finds seem to indicate the widespread presence of flat-bottomed design in the past.

The coracle is the likeliest source of the *umiak*. Alternatively, the kayak may have developed directly from the coracle—one that became narrow enough to lace the skin cover together on top. Transverse deck beams would keep the narrow hull from collapsing and the gunwales would become the principal strength members. A keel stringer would be developed longitudinally, crossed inside by transverse ribs. Other stringers outside would smooth the skin for the water flow. That kayak stringers usually end freely without joining onto the keel stringer may indicate development from the coracle independent of the *umiak*. The *umiak*'s distinguishing structural difference from the coracle is that on the bottom a pair of stringers join a central keel and/or end posts, if these are separate pieces. These two stringers are spread by transverse floor beams to form the flat bottom defined by hard chines from which the sides flare out. In Greenland, kayaks were essentially flat-bottomed until the last couple centuries when the V bottom—perhaps eight degrees or more dead rise amidships—developed on the west coast.

In historical times, only the eastern kayak family has a flat or near flat bottom with hard chines and flared, slablike sides. North Baffin and recent Foxe Basin–northwest Hudson Bay kayaks display a more graduated bottom-to-side transition, while the deeper V bottom prevails in much of West Greenland. The last is known to date only from the 17th century, while the multichine variant was limited to North Baffin until later in the 19th century. In summary, the East Canadian and Greenland hull design is flat-bottomed with hard chines at the bottom-to-side transition. To the west, all else is generally multichine with a round- or V-bottom hull and lacking that marked hard chine. Copper Inuit kayaks with just a single pair of stringers have hard chines and are deep V below. Some Nattilik (Netsilik) kayak bottoms seem near flat but the transition to the side is softened by extra stringers.

St. Lawrence Island Kayak Carvings. St. Lawrence Island is a large island in the Bering Sea measuring about a hundred miles (160 km) long and located 150 miles (240 km) south of Bering Strait. Several St. Lawrence midden sites located at the northwest tip near Gambell that were excavated in the early 1930s by Henry Collins (1937) have yielded kayak models of the Punuk culture, and possibly its Old Bering Sea ancestor. The sites are relatively dated by their beach line and distance from the receding sea. In 1995, John Heath visited the Smithsonian's National Museum of Natural History, in Washington, D.C., and examined the specimens with the assistance of Stephen Loring.

Apparently the oldest kayak in the Collins series is a solid wood model (Collins 1937:plate 59, fig. 1, NMNH A369828) from Miyowagh, which is the earliest site yielding kayak models. The artifacts from this site appear to be Old Bering Sea. The model (Figure 8.11) is plainly carved, symmetrical in plan with sharp ends. In profile, the ends are quite similar, with well-raked cutwaters.

Though level on top, they are high since the central three-quarters of the gunwales curve lower in a gentle sheer. Model length is 6⅛ inches (156 mm); the tips are somewhat worn. Beam measures one inch (24 mm), and amidships depth is ⅜ of an inch (10 mm). This model is distinct from the Ekven Cemetery models attributed to Okvik–Old Bering Sea and is difficult to interpret. The Miyowagh model appears intermediate both stylistically and in general cultural context between Okvik–Old Bering Sea and Punuk.

The boxy, flat-bottomed hull is first observed in two carved ivory models of classic Punuk culture (A.D. 700–1200), which continues into the historic period (Giddings 1967:155–157; Bandi 1969:78, 1995:167, 182). Both kayak models were indistinctly illustrated in profile, yet their moderately raked, angular ends suggest a flat-bottomed design (Collins 1937:plate 83, figs. 5 and 6, NMNH A355338). Recent examination of the two models by Heath and photographer Vernon Doucette confirms an *umiak*-like hull section with flat bottom, hard chines, and flared flat sides. The cockpit hoop is of the East Canadian Arctic D shape, with straight back tilted up in front. Recall that the historic hoop was minimally raised in front until the 20th century, perhaps when the waterproof gutskin kayak jacket was abandoned and the cockpit could no longer be sealed. The gunwales have a slight sheer or curve rising to the ends, while the bottom is moderately rockered.

Figure 8.11a—Prehistoric kayak model of solid wood found at Miyowagh on St. Lawrence Island. The model appears to be intermediate between Okvik–Old Bering Sea (300 B.C.–A.D. 500) and Punuk (A.D. 600–historic period). Length 156 mm. Photographer Vernon Doucette. National Museum of Natural History, no. A369828.

Figure 8.11b—View of the underside of the Miyowagh model. Note the shallowly rounded bottom. Photographer Vernon Doucette.

A small, carved model with a figure of a kayaker comes from the Punuk Ievoghiyoq site on St. Lawrence Island (Figure 8.12). Since its cockpit front is amidships while the opening is long, the kayaker is positioned well aft, again as in East Canadian Arctic kayaks. Greatest width is amidships, however, rather than at the back of the cockpit. The bow is a bit deeper than the stern, which is raked about 20 degrees out from vertical. The deck is moderately ridged down the middle. Ahead of the cockpit, a pair of transverse deck lines is engraved. Behind the kayaker is an unusual double float made of a pair of inflated sealskins. Such a float was used until recently in East Greenland for hunting large sea mammals (Holm 1914:47; Petersen 1986). At St. Lawrence, walrus and whale were prime quarries. The model is about 2⅜ inches (60 mm) long, ⁷/₁₆ of an inch (11 mm) wide, and ¼ of an inch (6 mm) deep to ridge.

Another carved model (NMNH A356213) from the old section of Gambell has the double float both fore and aft, the latter broken off to starboard (Figure 8.13). Although its tip is damaged, the stern is clearly wide on top. The deck surface has two depressions suggesting double cockpits, a small circular one amidships and a large D-shaped one aft that measures ⅜ inch (9 mm) long. The back of the larger depression, which is definitely a cockpit, comes far aft, at 2⅜ inches (61 mm) in the overall length of 3½ inches (89 mm) as measured with bits of both tips missing. A small depression is situated amidships and, if it is a cockpit, may be for a boy in training. The set of floats thus makes sense; the high-volume, wide stern bears the greater weight aft without too long a hull. Beam amidships is half an inch (12 mm), bottom width is ⅜ inch (10 mm), and depth to sheer is ⅜ of an inch (9 mm). The stern is slightly deeper than the bow. Both ends are more raked than in the single kayaker model, measuring about 35 degrees from vertical.

Especially intriguing is the well-carved ivory model from the Seklowaghyaget mound adjoining old Gambell on the side away from the sea and hence earlier (Collins 1937:plate 83, fig. 4, NMNH A364174). The model (Figure 8.14) was purchased from locals who were digging extensively around Gambell and so has no context. Seklowaghyaget is Punuk, a period when whaling was of great importance, more so than in the preceding Ievoghiyoq or Miyowagh sites (Collins 1937:186–187). The model recalls the historical Bering Strait kayaks of King Island, Diomede, and Seward Peninsula. The deep, near-vertical stern is wide on top; the gunwale rail ends protrude as in an *umiak*. The upcurving bow is pierced with a ¾-inch (20 mm) long slot in profile and in plan between the gunwale

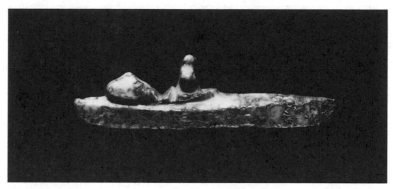

Figure 8.12a—Punuk ivory model with kayaker from Ievoghiyoq, St. Lawrence Island. Length 60 mm. Photographer Vernon Doucette. National Museum of Natural History, no. A355338.

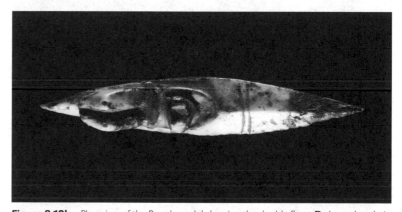

Figure 8.12b—Plan view of the Punuk model showing the double float, **D**-shaped cockpit, and paired foredeck lines. The double float also appears on an Old Bering Sea model from Ekven Cemetery, Chukotka, Siberia. Such floats were used for larger sea mammals, such as walrus. Photographer Vernon Doucette.

Figure 8.12c—View of Punuk model showing flat bottom, hard chines, slab sides, and attachment holes. The flat-bottom form first appears during the Classic Punuk period between A.D. 700 and 1200. Photographer Vernon Doucette.

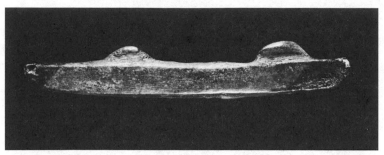

Figure 8.13a—Late Punuk ivory model of a double kayak from Old Gambell. Length 89 mm. Photographer Vernon Doucette. National Museum of Natural History, no. A356213.

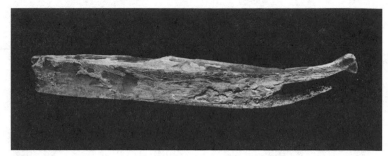

Figure 8.14a—Punuk ivory model kayak with pierced prow and ridged deck from the site of Seklowaghyaget. Length 120 mm. Photographer Vernon Doucette. National Museum of Natural History, no. A364174.

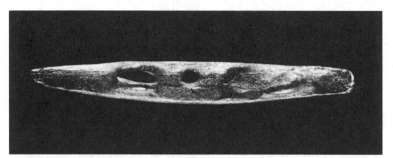

Figure 8.13b—Plan view of Late Punuk model showing contrasting cockpits, double float

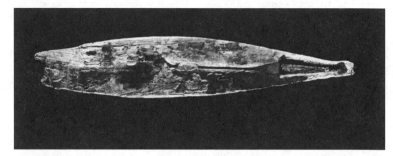

Figure 8.14b—Plan view of model from Seklowaghyaget showing wide stern with gunwale end protrusions and vertical piercing of bow. Photographer Vernon Doucette.

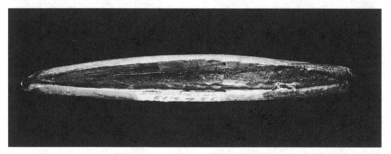

Figure 8.13c—View of Late Punuk model showing flat bottom with hard chines. Photographer Vernon Doucette.

Figure 8.14c—Bottom view of model from Seklowaghyaget. Photographer Vernon Doucette.

strips. If the slots are actual design features, the bow form is unique. Near the tip, the cutwater strip has a ⅜-inch (10 mm) gap due to breakage but was continuous originally. In profile, this bow suggests a possible source for the split form of the Aleutians and south Alaska kayaks, with width in the gunwales above the cutwater for throwing off waves and planing over them. Model length is 4¾ inches (119.5 mm), width is ⅞ inch (23 mm), depth sheer is ½ inch (13 mm), and depth ridge is ⅝ inch (17 mm). Beam at the front of the oval cockpit is 2⅝ inches (67 mm) from the stem, or front end.

With the cockpit back at 3³⁄₁₆ inches (81 mm), the kayaker sat well aft. Stern top width is half an inch (12 mm). The central hull cross section is rounded with flaring sides in King Island fashion, but the deck ridge is more moderate.

Another *umiak*-sterned ivory kayak model from St. Lawrence Island has just surfaced in private hands, unfortunately without site or context information and therefore undated. With a deep, raked stern such as those characterizing Aleut *umiat*, a plain, pointed bow rising in the Bering Straits manner, slightly bulged-up decks, and a

V bottom with hard chines, this example is a fascinating combination of features. The lack of contextual information is a great loss to kayak and maritime studies. The model would measure about 3⅞ inches (99 mm) long were the damaged bow tip intact. Width is ⅞ inch (21 mm) just before amidships, with a corresponding bottom width of ⅝ inch (17 mm), half an inch (13 mm) wide at the stern, and about the same depth overall with depth to sheer of about ⅜ inch (10 mm). The model is twice as large as the Ievoghiyoq specimen and polished with a lovely variegated patina.

The sides hollow in slightly under the gunwales at the stern as they do in *umiat* due to skin tension. The transverse suspension slot between the gunwales defines a thin crosspiece, but a wider one makes better structural sense and the slot location might be for convenience. Perhaps the model was worn as an amulet. The kayaker sat well back, as indicated by three peg holes aft of amidships. This design may be related to that of the Aleutian Islands, though they are about 600 miles (965 km) away. With the Asian mainland just some 50 miles (80 km) away and Alaska about 120 miles (193 km) from St. Lawrence Island, it is possible that kayak designs were shared across the Bering Sea (cf. Arima 2003).

A model *umiak* with a rising, pointed bow from Nunivak Island or the Kuskowim River delta region was collected in the 19th century. The model is complete with the *umialik*, or hunting captain, in the square stern, four crew, a seal aboard, and *palraiyuk* dragon painted on the side (Varjola 1988:32, fig. 41). The model's hard chines bring the Ievoghiyoq specimen to mind. If the two designs are related, this *umiak* model is intermediate in form between Ievoghiyoq and Seklowaghyaget. It is also similar to a wooden model from the Islands of the Four Mountains, located west of Umnak Island in the Aleutians. The model was recovered from a burial cave dating to the 14th century. Measuring 4 inches (about 10 cm) long, the model has a plain Bering Strait–like upcurving bow and a wide stern with a short vertical blade projection. It is the Aleut, or Unangan, *iqyax*, or kayak, design before the bow became bifurcate. The bladed stern form might be seen as an advance beyond the two wide-stern St. Lawrence designs, which may then be dated to the early second millennium A.D. Developmental and temporal orders need not coincide in the prehistoric design examples that are found, especially over great spatial and cultural distances, as in the case of the Aleutians and St. Lawrence Island. Kayak design transmission around the Bering Sea would not have occurred at a uniformly rapid rate.

Punuk-Thule and East Arctic Kayak Design. The Reindeer Chukchi skin-on-frame model illustrated by Bogoras (1904–1909:fig. 47b) has an elongated egg-shaped cockpit pointed at the front. Slightly widened tips seem to echo the wide, horizontally forked tips of the Ekven Cemetery models. A superb example published by Zimmerly (2000:3) demonstrates that valuable ancient models will continue to surface to gradually clarify our understanding of kayak development and arctic prehistory.

The Punuk culture derives its name from the Punuk Isles off the east end of St. Lawrence Island and is attributed to an influx from the Siberian mainland of warlike Eskimo using iron (Bandi 1995:168–171; Mason 1998:270). Although walrus have always been important resources in the region, whaling became an increasingly vital activity during Punuk times. Punuk is distinguished from the preceding Okvik and Old Bering Sea cultures by the presence of warfare. Bandi (1995:167) notes that "[w]hereas whaling certainly developed in the Bering Strait region in an Eskimo cultural context, warfare and its necessary equipment seems to have reached the Bering Strait region from Northeast Asia." He then reviews the evidence for Punuk warfare: body armor, sinew-backed bows and lethal-looking arrows on St. Lawrence Island, wrist guards, the skeleton of a man who had been riddled with arrows, and high lookout and defensive house sites (Bandi 1995:168–180). Punuk may be the source of the flat-bottomed, hard-chine kayak design. Bandi suggests that whaling fostered warfare:

> Raids…certainly demanded some kind of leadership. The same is true of whaling. The increase of this most dangerous variant of sea mammal hunting during the Punuk period may have led to a perfection of navigation techniques, improved umiaks, disciplined crews, and boat captains with strict military-like authority. Even today on St. Lawrence Island, a boat captain, who on shore might be a reserved man, has absolute command on the umiak, during the hunt, and in the distribution of captured whales, although he has to respect traditions. It is obvious that successful attacks on other villages, whether carried out on land or by boat, depended upon good leadership.…

> Punuk period villages are larger and more numerous than are those of the preceding Old Bering Strait occupation, suggesting that regional population density was greater and communities were more complexly organized. Sociopolitical organization is similar to the ethnohistoric period, would

have been based on large, extended families headed by umialiqs or production leaders who served as "big men." Maritime subsistence, thought to have been based primarily on walrus and baleen whales, relatively high population density, needs for intersocietal alliances, control of transcontinental trade, and new weaponry/armor technology from distant Asian peoples, seems to have combined in the Bering Strait region to provide a cultural environment for persistent, large-scale armed conflict (Bandi 1995:180–181).

These numerous, bold, and on-the-move Punuk whalers seem to be natural candidates for the vanguard of the Thule culture expansion eastward from northern Alaska in the 13th century A.D. (McGhee 2000). Although Thule culture is usually regarded as developing from Birnirk, it has Punuk elements, such as the Sicco open-socket harpoon head found in Alaska, the East Canadian Arctic, and Greenland (Collins 1937:119, 183, 203), and the flat-bottomed, hard-chine kayak design evident in the ivory model from Ievoghiyoq.

A number of carved wooden Thule kayak models have been found, mostly from the Canadian Arctic islands. Most are round-bottomed with curved raked cutwaters, suggesting an affinity with the ancestral Birnirk kayak and the later historic Caribou, Natsilik (Netsilik), Copper, and Mackenzie Inuit kayaks. A few Thule flat-bottomed models with straight raked cutwaters do occur in Greenland and the Canadian Arctic. An outstanding specimen is a model from the Clachan site (Figure 8.15) on the west end of Coronation Gulf, which is dated roughly to the 14th century A.D. (Morrison 1983:167, CMC NaPi-2-29-15). The Clachan model is very much an East Canadian Arctic design, with the big high bow that is as much for windage as for surmounting waves. With a wide, shallow-draft flat-bottomed hull, the stern tends to blow around too easily downwind unless the bow catches more wind to counterbalance. Perhaps the original Punuk design was found lacking in this respect. In Greenland, narrow, low-volume hulls developed so the bow is more moderate.

Based on the form and size of five paddles found in northeast Greenland in the early 1930s by Søren Richter (1934:142–143, fig. 78), the Punuk-Thule type kayak may have reached the easternmost extreme of Thule expansion. Recovered from various places in Eirik Raudes Land, with a single exception from the Danish National Museum (catalog number 33042), the paddles (Figure 8.16) have long, narrow blades separated by collar expansions from the grip section. The one nearly complete specimen (Danish National Museum, cat. no. 33048), found on the outer coast of Gauss, is over 11 ½ feet

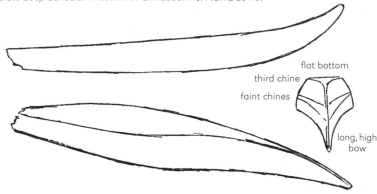

Figure 8.15—Thule Culture East Arctic kayak wooden model, 1.92 mm minus stern tip, found on west Coronation Gulf at the Clachan site, A.D. 1150–1450. This model provides important early evidence of the East Canadian Arctic design (Morrison 1983:167, 347, plate 26c). Canadian Museum of Civilization no. NaPi-2-29-15.

Figure 8.16—Prehistoric northeast Greenland paddles and partial reconstruction of a kayak framework. The shapes are East Canadian Arctic and predate historically known Greenland forms (after Richter 1934:142, fig. 77; 143, fig. 78). Danish National Museum. Catalog numbers are listed to the right of each paddle. Drawn to different scales.

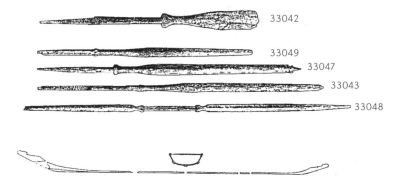

(3.5 m) long, which suggests a kayak measuring over 23 feet (7 m), according to the tradition of making this kind of paddle half the length of the kayak. A tip is broken, but calculated from the whole blade and grip the paddle is about 13¼ feet (4 m) long, as great a size as is known historically for the East Canadian Arctic and implying a 26-foot (8 m) class kayak, the largest voyaging model. Such a large kayak contradicts a previous interpretation based on European historical records that East Arctic kayaks, at least in Canada, were generally only in the 13- to 17-foot (4 to 5 m) range until as late as the 18th or 19th century (Arima 1987:78–81, 97).

Similar calculations applied to three other paddles, which are halves of originals scarfed together in the middle, yield kayak lengths of 16½ feet (5 m, based on specimen 33049) to 20 feet (6 m, based on paddles 33047 and 33043). Calculations based on the short paddle (33042) yield a 12-foot (3.7 m) kayak, perhaps a floe edge retriever or a boy's model. Its blade, from the grip section out, measures 19 by 3¾ inches (48 by 9.5 cm) and is much stubbier than the others, which measure between 40 and 67 inches (100 to 170 cm) long by 2⅜ to 2¾ inches (6 to 7 cm) wide, tapering to the end. Four similar paddles, all halves, were found at Cape Hedlund. The blades measure up to 4 feet 7 inches (1.4 m) long and 3⅛ inches (8 cm) wide (Glob 1935:70). Extrapolation suggests a 22-foot (3.8 m) kayak, an ideal size in the East Canadian Arctic.

A kayak framework in pieces was also found by Richter on the south side of the entrance to Segelsällskapets Fjord off King Oscar Fjord (Petersen 1986:142–143). Since part of the hull is missing, length cannot be determined, but it exceeds 16½ feet (5 m). Maximum width is estimated to be at least 25½ inches (65 cm) from the longest crosspiece, which is 24½ inches (62 cm) long. Such a wide beam is typical of East Canadian Arctic kayaks. According to the maximum cross section shown in Richter's partial reconstruction (1934:142, fig. 77), the hull is also deep. The depth is suggestive of the East Canadian Arctic style, which is usually deeper than that of Greenland craft, at least since the 18th century.

Since the longest deck beam is straight, the cross section would be from just behind the cockpit; thus, the maximum width is shifted aft as in East Canadian Arctic kayaks. Cockpit front beams are arched. Richter's reconstruction does look like some historical northern West Greenland kayaks—it has a shallow V bottom and the bow is higher than the stern. The high bow–low stern combination is typically East Canadian Arctic. Characteristically Greenlandic are the keel ends that expand vertically into deep cutwater sections. Assuming that the V-bottom reconstruction is correct, the kayak seems to be just what one would expect for an early Greenland kayak exhibiting strong East Canadian Arctic influence, particularly the wide width.

Dorset Culture Fragments in North Baffinland. Any study of design origins must consider whether the Dorset culture of about 800 B.C. to A.D. 1000–1400/1500 had watercraft and, if so, during what time period. The usual terminal date given by archaeologists for Dorset is A.D. 1000. In some areas, Dorset appears to have survived later, overlapping with the Thule culture until A.D. 1400/1500 (Maxwell 1984; McGhee 1984; Sutherland 2000:164).

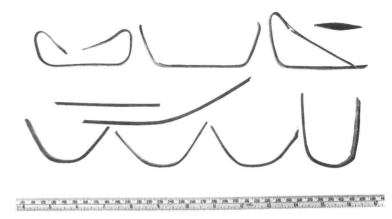

Figure 8.17—Late Dorset model watercraft pieces from North Baffin. Top right: kayak shape from Nunguvik; length 47 mm. Top and middle rows: flat bottom ribs and possible keel sections from Nunguvik. Bottom row: Button Point rounded ribs. Photographer E. Arima. © Canadian Museum of Civilization, nos. PgHb-1:2451 and PfFm 1:246.

Although the coastal and island locations of most Dorset culture sites indicate a maritime adaptation and therefore the presence of watercraft, arctic archaeologists, particularly Canadian ones, have debated the existence of Dorset skin boats because of the lack of evidence. The strongest argument for Dorset watercraft comes from North Baffin in the possible model and actual kayak and/or *umiak* ribs (Figures 8.17, 8.18) recovered in the Mittimatalik or Pond Inlet region by Mary-Rousselière (1979). Since these Dorset finds are vital to the debate, his description of the evidence from the Nunguvik site on west Navy Board Inlet will be quoted in full:

> Parts of kayak or umiak models had already been found at Button Point in 1962 [Figure 8.19], but in dubious stratigraphy. In 1970, several similar pieces of a kayak model were discovered in house 71. They were ribs…straight at the bottom and meeting on the deck. Then, in 1974, a flat piece of wood, from 4 to 5 mm thick, measuring 312 mm by 30 to 40 mm and bent at an angle of 136 degrees was found in house 76. One end had been broken just after bending at a similar angle; the other was slightly bevelled, and the corresponding extremity of another wooden lathe was still in place. This could hardly be anything but a rib fragment. The piece had been bent after being partly cut across on both faces, and tiny holes were regularly spaced on both sides of the piece, for the lashing of the longitudinal slats with sinew

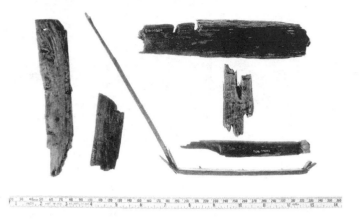

Figure 8.18—Dorset wood fragments from the Nunguvik site. The three pieces on the left were previously identified as kayak ribs but are probably hexagonal box strips. Right, top to bottom: possible gunwale, stringer, and rib fragments. Photographer E. Arima. © Canadian Museum of Civilization, nos. PgHb-1:4032, PgHb-1:11477, PgHb-1:4026, PgHb-1:11520, PgHb-1:4229, PgHb-1:4293.

Figure 8.19—Button Point Dorset possible end top crosspiece for small *umiak* recessed for midline vertical member, gunwales, and perhaps flotation bladders; estimated length 20 cm. Photographer: E. Arima. © Canadian Museum of Civilization, no. PfFm-1:293.

thread, fragments of which were still in the holes. The whole rib was carefully smoothed on the outside.

These finds support each other and suggest that the Dorset kayak was not very different from the modern sea kayak of the Baffinlanders, the main difference being that the ribs were more angularly bent, instead of being curved. Both the deck and the bottom appear to have been flat (Mary-Rousselière 1979:25–26).

The modern North Baffin kayak familiar to Mary-Rousselière has curved ribs and a multichine section as noted above. Elsewhere in the East Canadian Arctic, ribs are more angular; at times three straight sections are joined together, giving the characteristic flat bottom with hard chines and slablike flared sides. Moreover, the degree of bend in the Nunguvik specimen (PgHb-1:4026) is typical for modern East Canadian ribs, at least as illustrated at 116 degrees, not 136 degrees as mentioned in the text, which is probably an error. Currently the angle is 122 degrees, doubtless from relaxation.

The article illustration of the rib (Mary-Rousselière 1979:26, fig. 5) shows the piece on its side. This section is 7 ¼ inches (18 cm) long, while the other arm is 5 ¼ inches (13 cm). The thin lathelike strip is nearly 1 ¼ inches (3 cm) wide; although one edge has broken away, the strip is probably close to its original size, since shorter but comparable fragments with both edges intact are of similar width (e.g., PgHb-1:4032, 1:4229). Thickness is about ¼ inch (6 mm).

Two short lathelike fragments (also numbered PgHb-1:4229, this accession number includes five pieces) appear to be stringer bits. One is 7⁄16 inch (1 cm) thick by 1 13⁄16 inches (3 cm) wide. The other has a broken surface, so its original thickness is unknown. It is a particularly significant fragment, however, at 1 ¼ inches (3 cm) wide by 3 inches (8 cm) long; at each broken end is part of a slitlike hole cut lengthwise but with the pair staggered slightly across the grain as in historical kayak construction for a framework tie. The space between the holes is 1 3⁄8 inches (3.5 cm), suitable for tying the piece onto a rib. The split-away face may have been the outer one with the lashing countersink groove, which fostered breakage. Remaining thickness is 5⁄16 inch (8 mm). The fragment may be a stringer or light keel strip.

The presumed rib piece displays a couple of surprising details. The top of the side section was attached to a 2 3⁄8 inches (6 cm) long lathelike fragment. A sinew lashing hole through both pieces confirms that the association was not accidental. Both pieces are tapered in thickness for this scarf joint. In storage they have become separated, and only recently (July 1995) were they reassembled for closer examination.

Three fine lashing holes are located in the side section of the possible rib about ¼ inch (6 mm) from the edge and at 1, 3, and 4 ¾ inches (25, 75, 120 mm) from the bend. In the bottom section are three more holes, a middle one slightly off center for the keel stringer and one to each side about 1 inch (2.5 cm) in from the corners. Fine countersink grooves connect the holes to the edge to house the sinew lashing, a few bits of which remained in the holes when the piece was recovered. One of the stringerlike fragments mentioned above (PgHb-1:4229) has staggered holes to run the lashing across the grain, a principle observed in Inuit woodworking.

Another puzzling detail of this riblike piece is that the intact edge is stained red. Two spots on the inside surface of the bottom section and another on the side inside surface near the bend are also red. Blood can be used as glue, yet it is unusual for a rib to be glued on edge. The red color may be related to symbolic or decorative purposes. The edge staining does not support interpretation as a rib and hence of a Dorset kayak.

The 2 ⅓-inch (6 cm) section found against the inner face of the intact arm of the supposed rib is tapered thin for a scarf slanted through the thickness. When the pieces are assembled, with clean faces together and the tiny sinew tie hole in each member aligned, the short section's thicker end is at the end of the whole arm of the large piece, as illustrated by Mary-Rousselière (1979). It may be a broken kerfed bend at an angle like those in the larger piece. Together with a third section, these pieces may join to form the bottom side strip of an elongated six-sided box, with a couple more strips added above to create more depth. Such a box would account for the sinew lashing holes near the edge and the red staining, perhaps from blueberry or stoveberry juice. A right-angled bend, too prone to breaking, may explain why it is six-sided rather than four-sided. Such an unusual shape can be related to the supposed Viking influence discerned by Patricia Sutherland in the remains of house 71, for example, spun yarn, European-style woodworking, and model skis (Sutherland 2000:160–163; Mary-Rousselière 2002:115–116).

The model kayak and/or *umiak* ribs of baleen found by Mary-Rousselière do not prove that full-size Dorset watercraft existed. But Jarmo Kankaanpää (1996) has argued that the existence of Dorset kayaks is certain based on artifactual evidence and the physical circumstances of maritime existence. For Kankaanpää, Dorset sea kayaking was so well developed that the Thule Inuit adopted it to upgrade their limited kayaking in sheltered waters, as in North Alaska. Thule kayaking was secondary to *umiak* hunting for Kankaanpää, who is thinking in terms of the tippy multichine kayak traced to the Birnirk culture of northern Alaska. But the central artifact of the East Canadian Arctic sea kayak itself is the stable, flat-bottomed hull form of the Punuk culture. We must await further evidence to decide the Dorset kayak and *umiak* question. In the comparative study of kayaks, however, a Dorset design is not needed to account for extant forms—the Punuk Thule and Birnirk Thule designs suffice.

REFERENCES

Adney, Edwin T., and Howard I. Chapelle
[1964] 1983 *The Bark Canoes and Skin Boats of North America*. Washington, D.C.: Smithsonian Institution Press.

Alaralak, Felix
1990 Interview. Inullariit Society Archives IE-114. Igloolik, Nunavut.

Apakok, Ali
1985 From the Belchers to Great Whale by Kayak. *Inuktitut* 60: 4–18.

Arima, Eugene Y.
1964 Notes on the Kayak and its Equipment at Ivuyivik, P.Q., pp. 221–261 in *Contributions to Anthropology, 1961–62*, Part II. Ottawa: National Museum of Canada Bulletin 194.
1975 *A Contextual Study of the Caribou Eskimo Kayak*. Ottawa: National Museum of Man, Canadian Ethnology Service, Mercury Series no. 25.
1980 The Tuniit–Inuit Connection. *Inuktitut* 47:43–52.
1985 Qainnat Origins. *Inuktitut* 60:19–34.
1987 *Inuit Kayaks in Canada: A Review of Historical Records and Construction*. Ottawa: Canadian Museum of Civilization, Canadian Ethnology Service, Mercury Series no. 110.
1989 Speculating on Skin Boats of Antiquity: Coracles, Umiaks, Kayaks, Bark Canoes, and Origins. Paper presented at the Skin Boats of Antiquity Conference, Kodiak, Alaska.
1991 Revival in Canada: A Bicultural Presentation, pp. 107–130 in John D. Heath et al., *Contributions to Kayak Studies*. Ottawa: Canadian Museum of Civilization Mercury Series, Canadian Ethnology Service, Paper no. 122.
2003 Grail I and Grail Too. *Qajaq* 1:6–10.

Arutyunov, S. A., M. G. Levin, and D. A. Sergeyev
1964 Ancient Burials on the Chukchi Peninsula, pp. 319–326 in Henry N. Michael, ed., *The Archaeology and Geomorphology of Northern Asia: Selected Works*. Anthropology of the North, Translations from Russian Sources 5. Toronto: University of Toronto Press.

Bandi, Hans-Georg
1969 *Eskimo Prehistory*. Trans. Ann E. Keep. Fairbanks: University of Alaska Press.
1995 Siberian Eskimos as Whalers and Warriors, pp. 165–183 in A. P. McCartney, ed., *Hunting the Largest Animals: Native Whaling in the Western Arctic and Subarctic*, Studies in Whaling 3, Occasional Paper 36, Circumpolar Institute, University of Alberta, Edmonton.

Bernier, James E.
1910 *Cruise of the* Arctic *1908–09: Report on the Dominion of Canada Expedition to the Arctic Islands and Hudson Strait on Board the D.G.S.* Arctic. Ottawa: Government Printing Bureau.

Boas, Franz
1888 *The Central Eskimo*. Washington, D.C.: Smithsonian Institution, Bureau of American Ethnology Annual Report 6 (1884–85): 399–666.
1901–1907 *The Eskimo of Baffin Land and Hudson Bay: From Notes Collected by Captain George Comer, Captain James Mutch, and Reverend E. J. Peck*. New York: American Museum of Natural History Bulletin XV.

Bogoras, Waldemar
1904–1909 *The Chukchee Jesup North Pacific Expedition*, vol. VIII, part 1. Memoir of the American Museum of Natural History XI. Leiden and New York.

Brand, John
1984 *The Little Kayak Book*, Part I. Colchester, U.K.: John Brand.

Collins, Henry B.
1937 *Archaeology of St. Lawrence Island, Alaska*. Washington, D.C.: Smithsonian Institution Miscellaneous Collections 96(1).

Damas, David J., ed.
1984 *Arctic*, Handbook of North American Indians, vol. 5. Washington, D.C.: Smithsonian Institution.

Flaherty, Robert J.
1924 *My Eskimo Friends, "Nanook of the North."* Garden City, NY: Doubleday, Page and Co.
1938 *The Captain's Chair: A Study of the North*. London: Hodder and Stoughton.

Freeman, Milton M. R.
1964 Observations on the Kayak-Complex, Belcher Islands, N.W.T., pp. 56–91 in *Contributions to Anthropology, 1961–62*, Part II. Ottawa: National Museum of Canada Bulletin 194.

Giddings, James Louis
1967 *Ancient Men of the Arctic*. New York: Alfred A. Knopf.

Gilberg, Rolf
1974–1975 Changes in the Life of the Polar Eskimos Resulting from a Canadian Immigration into the Thule District, North Greenland, in the 1860s. *Folk* 16 & 17:159–170.

1984 Polar Eskimo, pp. 577–594 in David J. Damas, ed., *Arctic*, Handbook of North American Indians, vol. 5. Washington, D.C.: Smithsonian Institution.

Glob, P. V.
1935 Eskimo Settlements in Kempe Fjord and King Oscar Fjord. *Meddelelser om Grønland* (Copenhagen) 102(2).

Grønnow, Bjarne
1988 Prehistory in Permafrost: Investigations at the Saqqaq Site, Qeqertasussuk, Disco Bay, West Greenland. *Journal of Danish Archaeology* 7:24–39.

Grønnow, Bjarne, and Morten Meldgaard
1988 Boplads i dybfrost–Fra Chistianshåb Museums udgravninger på Vestgrønlands ældste boplads. *Naturens Verden* 1988 (11 & 12): 409–440.

Guemple, D. Lee
1966 The Pacalik Kayak of the Belcher Islands, pp. 152–218 in *Contributions to Anthropology, 1963–64*, Part II. Ottawa: National Museum of Canada Bulletin 204.

Heath, John D.
1991 The King Island Kayak, pp. 1–38 in Heath et al., *Contributions to Kayak Studies*. Ottawa: Canadian Museum of Civilization Mercury Series, Canadian Ethnology Service Paper 122.

Heath, John D. et al.
1991 *Contributions to Kayak Studies*. Ottawa: Canadian Museum of Civilization Mercury Series, Canadian Ethnology Service Paper 122.

Hendrik, Hans
1878 *Memoirs of Hans Hendrik, the Arctic Traveler, Serving under Kane, Hayes, Halland Nares, 1853–1876*. Trans. Henry Rink. London: Trübner.

Holm, Gustav
1914 Ethnological Sketch of the Angmagsalik Eskimo, pp. 225–305 in William Thalbitzer, ed., *The Ammassalik Eskimo*, part 1. *Meddelelser om Grønland* (Copenhagen) 39.

Holtved, Erik
1967 Contributions to Polar Eskimo ethnography. *Meddelelser om Grønland* (Copenhagen) 182(2).

Kane, Elisha Kent
1856 *Arctic Explorations: The Second Grinnell Expedition in Search of Sir John Franklin 1853, 1854, 1855*. Philadelphia: Childs and Peterson.

Kankaanpää, Jarmo Kaleri
1996 *Thule Subsistence*. Ph.D. dissertation, Department of Anthropology, Brown University, Providence, RI.

Lee, Molly, and Gregory A. Reinhardt
2003 *Eskimo Architecture: Dwelling and Structure in the Early Historic Period*. Fairbanks: University of Alaska Press.

Lofthouse, Joseph
1922 *A Thousand Miles from a Post Office, or, Twenty Years Life and Travel in the Hudson's Bay Regions*. Toronto: Macmillan.

Lyon, George F.
1824 *The Private Journal of Captain G.F. Lyon of H.M.S.* Hecla, *during the Recent Voyage of Discovery under Captain Parry*. London: John Murray.

Mary-Rousselière, Guy
1979 A Few Problems Elucidated…and New Questions Raised by Recent Dorset Finds in the North Baffin Island Region. *Arctic* 32(1):22–23.
1980 *Qitdlarssuaq: l'histoire d'une migration polaire*. Montreal: Presses de l'université de Montréal.
1991 Report on the Construction of a Kayak at Pond Inlet in 1973, pp. 39–78 in John D. Heath et al., *Contributions to Kayak Studies*. Ottawa: Canadian Museum of Civilization Mercury Series, Canadian Ethnology Service Paper 122.
2002 *Nunguvik et Saatut: Sites paléoeskimaux de Navy Board Inlet, île de Baffin*. Ottawa: Canadian Museum of Civilization Mercury Series, Archaeological Survey of Canada Paper 162.

Mason, Owen K.
1998 The Contest Between the Ipiutak, Old Bering Sea, and Birnirk Polities and the Origin of Whaling during the First Millennium A.D. along Bering Strait. *Journal of Anthropological Archaeology* 17(3):240–325.

Mathiassen, Therkel
1927 Archaeology of the Central Eskimos. Fifth Thule Expedition 1921–24 Report 4(1 & 2). Copenhagen: Gyldendalske.
1928 Material Culture of the Iglulik Eskimos. Fifth Thule Expedition 1921–24 Report 6(1). Copenhagen: Gyldendalske.

Maxwell, Moreau S.
1984 Pre-Dorset and Dorset Prehistory of Canada, pp. 359–368 in David J. Damas, ed., *Arctic*, Handbook of North American Indians, vol. 5. Washington, D.C.: Smithsonian Institution.

McClintok, Francis Leopold
1860 *The Voyage of the* Fox *in Arctic Seas*. Philadelphia: V. T. Lloyd.

McGhee, Robert
1984 Thule Prehistory of Canada, pp. 369–376 in David J. Damas, ed., *Arctic,* Handbook of North American Indians, vol. 5. Washington, D.C.: Smithsonian Institution.
2000 Radiocarbon Dating and the Timing of the Thule Migration, pp. 181–191 in Martin Appelt, Joel Berglund, and Hans Christan Gulløv, eds. *Identities and Cultural Contacts in the Arctic*. Copenhagen: Danish National Museum and the Danish Polar Center.

Morrison, David A.
1983 *Thule Culture in Western Coronation Gulf, N.W.T.* Ottawa: National Museum of Man Mercury Series, Archaeological Survey of Canada Paper No. 116.

Nasook, Martha
1991 Interview. Inullariit Society Archives IE-175. Igloolik, Nunavut.

Parry, William E.
1821 *Journal of a Voyage for the Discovery of a North-West Passage from the Atlantic to the Pacific; Performed in the Years 1819–20, in His Majesty's Ships* Hecla *and* Griper. London: John Murray.
1824 *Journal of a Second Journey for the Discovery of a North-West Passage from the Atlantic to the Pacific; Performed in the Years 1821–22–23 in His Majesty's Ships* Fury *and* Hecla, vol. 1 of 2. London: John Murray.

Petersen, H. C.
1986 *Skinboats of Greenland*. Ships and Boats of the North, vol. 1. Roskilde: Trans. Katharine M. Gerould. The National Museum of Denmark, the Museum of Greenland, and the Viking Ship Museum in Roskilde.

Piugaattuk, Noah
1986, 1990, 1991, 1993, 1994 Interviews. Inullariit Society Archives IE-202, -136, -181, -248, -303. Igloolik, Nunavut.

Rae, John
1850 *Narrative of an Expedition to the Shores of the Arctic Sea in 1846 and 1847*. London: T. and W. Boone.

Rasmussen, Knud
1908 *The People of the Polar North*, G. Herring, ed. London: Kegan Paul, Trench, Trubner.

Richter, Søren
1934 A Contribution to the Archaeology of North-East Greenland. *Etnografiske Museums Skrifter* (Oslo) 5(3):65–216. Oslo: A.W. Brøggers Boktykkeri.

Ross, William Gillies
1975 *Whaling and Eskimos: Hudson Bay 1860–1915*. Ottawa: National Museums of Canada, National Museum of Man Publications in Ethnology 10.

Silis, Ivars
1984 Narwhal hunters of Greenland. *National Geographic* 165(4): 520–539.

Sutherland, Patricia D.
2000 Strands of Culture Contact: Dorset-Norse Interactions in the Canadian East Arctic, pp. 159–169 in M. Appelt, J. Berglund, and H. C. Gulløv, eds. *Identities and Cultural Contacts in the Arctic*. Copenhagen: Danish National Museum and the Danish Polar Center.

Turner, Lucien M.
1894 Ethnology of the Ungava District, Hudson Bay Territory. Washington, D.C.: Smithsonian Institution, Bureau of American Ethnology Annual Report 11 (1889–90):159–350.

Tyrrell, Joseph B.
1898 *Report of the Doobaunt, Kazan and Ferguson Rivers and the North-West Coast of Hudson Bay*. Ottawa: Canadian Geological Survey Annual Report n.s. 9(F).

Uyarasuk, Titus
1990, 1991 Interviews. Inullariit Society Archives IE-110 & -179. Igloolik, Nunavut.

Varjola, Pirjo
1988 *Alaska: Venäjän Amerikka/Russian America*. Helsinki: National Board of Antiquities.

Wilkinson, Douglas
1955 *Land of the Long Day*. New York: Henry Holt & Co.

Zimmerly, David W.
2000 *Qayaq: Kayaks of Siberia and Alaska*, 2nd ed. Fairbanks: University of Alaska Press.

Glossary

afterdeck — The area of the deck between the aft side of the cockpit hoop and the stern.

aftermost — Nearest the stern or back end of the boat; most aft.

aftermost edge of the cockpit — The portion of the edge of the cockpit rim closest to the stern.

amidships — Midway between the bow and the stern of the kayak, or midway from side to side.

angle of attack — The angle between the wide plane of a paddle blade and its direction of travel. Hold your hand or some flat object out of a car window. Start the experiment at low speeds until you know how much force might be generated. If the wide axis of your hand or the flat object is parallel to the wind direction, there is no lift. As you begin to raise or lower the leading edge, a vertical force is generated. As the angle of attack increases, the vertical force increases, but only up to a point. With too much angle of attack, the vertical force is reduced or may even become zero, while the drag or effort to move your hand or the paddle increases. The kayaker needs to learn what the optimum angle of attack is.

beam — The width of a boat.

beam ends — When the kayak is laid on its side, it is said to be on its beam ends (the deck beams are standing on end) and thus in danger of capsizing.

beam seas — Waves coming from the side of the kayak.

bilge stringers — The lightweight lengthwise structural members that run along the turn of the bilge or lower rounded part of the kayak hull.

blade — The wide, flat portion of the paddle that provides the push. See diagram, page 151.

bow — The forward end of the kayak.

brace — A maneuver in which the paddle (and sometimes the submerged torso) acts as a support to keep the kayak upright. This often involves **sculling**.

capsize — Overturning the kayak or other vessel.

chine — A feature of the hull; also defined as the intersection of the bottom and sides of a flat or V-bottomed boat. Seen from either end, a "hard chined" boat has an abrupt intersection; a "soft chined" boat has a gradual change. A round-bottomed boat has little or no chine. A boat may have several chines. See diagram, page 150.

cockpit rim — The wooden hoop encircling the cockpit.

deck — The top of the kayak extending from the bow to the stern.

deck beam — The wooden cross beams that generally run horizontally across the kayak between the **gunwales**. They are also called **thwarts**.

deep gunwale — The wide vertical dimension of the piece of wood that is the **gunwale**.

"floating" — Condition in which the cockpit rim or hoop is not rigidly or structurally attached to the rest of the kayak framework, but rather is only attached to the overlying skin of the kayak. This allows it to move slightly or "float."

foredeck — Area of the deck between the cockpit hoop and the bow.

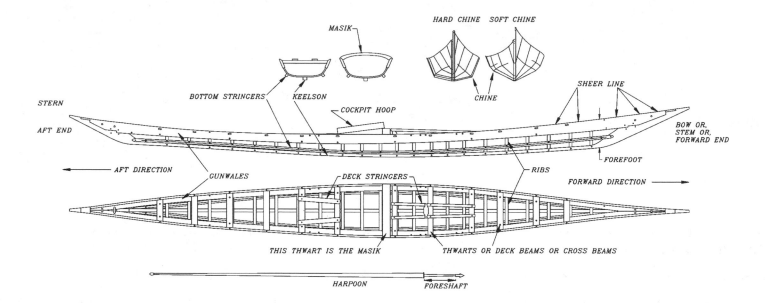

forefoot — The part of the kayak from the deck to the place where the **keelson** meets the **stem**.

foremost — Forwardmost or toward the bow.

foreshaft — A piece of bone or walrus tusk at the front of the harpoon that acts as a failsafe feature if the prey struggles violently. It is lashed to the front of the wooden harpoon in such a way that all of the impact forces striking the prey are transmitted to the harpoon point. However, if the prey exerts a strong side force that might break the harpoon shaft, the shaft dislocates at the junction of the foreshaft and the main shaft. The foreshaft also adds weight to the harpoon to move the center of balance forward to improve accuracy and penetration.

gunwales — Upper edge of the kayak where the deck meets the side of the hull. Also, the principal wooden boards running the length of the kayak at the gunwales.

heeled — The inclined position of the kayak to one side or the other.

hogging — Condition in which the ends of a boat droop and the mid-section rises.

hydrofoil — Winglike structure used in water.

inboard — The direction inward toward the longitudinal centerline of the kayak; opposite of outboard.

keelson — Longitudinal structural member that runs along the bottom of the hull inside the sealskin covering. In nautical terminology, the keel is an exterior member and the keelson is interior.

longitudinal stringers — Lightweight, lengthwise structural member of the kayak frame. Usually does not include the **keelson** or **gunwales**.

longitudinals — Parts of the kayak that run lengthwise. Usually refers to the **keelson** and the other long thin strips of wood running along the length of the bottom of the kayak, but may be used to refer to the deck **stringers**.

loom — The shaft of the paddle near the center. See diagram, page 151.

masik — Greenlandic term for the wide, curved main deck beam that also supports the forward edge of the cockpit hoop.

outboard — The direction outward from the kayak's side or laterally away from the longitudinal centerline of the kayak; opposite of inboard.

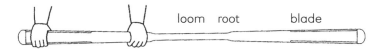

loom   root   blade

outrigger—Typically, a harpoon, piece of wood or paddle rigged out from the side of the kayak to prevent capsizing.

paatit—Greenlandic for paddle. Other possible former spellings include *pautik*.

port—The left side of the kayak when facing the bow.

raked cockpit—Condition in which the cockpit hoop is tilted toward the bow; thus it is highest at the foremost edge and lowest at the aftmost edge.

ribs—Curved, transverse timbers or members forming the kayak frame.

root of blade—Area where the blade narrows down to the **loom**.

sagging—When the ends of a boat rise and the midsection droops.

scull—To oscillate the paddle blade back and forth in the water to generate lift.

sheer or sheer line—The upward curve of the line of the kayak's gunwale as seen from the side; usually refers to the curve near the bow, but may also refer to the curve or lack of it along the entire length of the kayak. "Reverse sheer" is the condition in which the gunwale is higher near the middle and lower at the ends.

skeg—Small finlike attachment at the aft end of the keelson that helps to make the kayak more directionally stable. Proper kayak design is a balance between directional stability and maneuverability.

stall—When a paddle has too much angle of attack, it loses lift and stalls.

starboard—The right side of the kayak when facing the bow.

stern—Back end of the kayak.

stem—The vertical structural member at the bow of the kayak to which the **gunwales** and **keelson** are attached.

storm paddle—The use of the phrase "storm paddle" in this text only refers the short paddles common in regions with bad weather. Generally Greenlanders did not carry spare paddles.

stringers—The lengthwise structural members of the kayak framework, usually other than the **gunwales** and **keelson**.

swells—Long rolling waves that do not break; also called rollers.

thwarts—The cross beams or **deck beams** of the kayak.

Bent Frederiksen of Nuussuaq, Upernavik, throwing his *unaaq*, or harpoon, August, 1968.
Courtesy Keld Hansen.

Index

Page numbers in **boldface** refer to diagrams, maps, and photographs.

A

aariammillugu (paddle held lengthwise along spine, hands held fore and aft), 35
Aberdeen, Scotland, grounded Greenland kayak (1824), **10**, 13–14
Absalosen, Kamp, 58
afterdeck, defined, 149
Aggumiut multichine kayak design, 124
Agu Bay, Baffin Island, 123
Ainalik (Ivuyivik), 125, **126**, **130**
Aivilingimiut (Igloolik) kayak designs, 118–122, 123
   See also Igloolik oral history
Akudnirmiut kayak designs, 124, 127
Akuliaruseq, Greenland, 100–101
Åland kayak, 79
Alaralak, Felix, 136
Alaska, 7, 20–21
   See also St. Lawrence Island
Aleutian Islands
   baidarka, v, 88
   prehistoric kayak designs, 141
American Museum of Natural History, New York, 5, 7
amidships, defined, 149
Amphibious Man (video), 49
angle of attack, defined, 149
anorersuarmi kinngusaqattaarneq (capsize maneuvers storm techniques), 31–33, **32**
anthropomorphic sizing
   of Greenland paddles, 14, 47–48, 49, 131
   of kayaks, 131
Arctic, North American, **viii** (map)
arviq/aqviq (bowhead or Greenland whale; *Balaena mysticetus*), 117–118, 133, 135
assak peqillugu, qilerlugu/poorlugu (clenched fist roll), 39
assammik kingukkut (hand only, leaning aft), 39
assammik masikkut (hand only, sweep outward and down from near *masik*), 39, **40**

assammik nerfallaallugu (hand only, sweep aft, then forward), 39
Ataralaa's kayak trip (1899–1900), 103–105
avasisaartoq (curved kayak), 65–66, **66**, 67
avataq (hunting or sealskin float). *See* sealskin floats (*avataq*)
avataq roll (roll using sealskin float), 37–38, **37–38**

B

Baffin Island, **108** (map)
Baffin Island kayak, 85, **86**, 90
   kayak design history, 123–125, 143–144, **143–144**
   See also Igloolik oral history; North Baffin Island; Polar Inuit (Greenland)
baidarka, v, 88
balance training
   for adults, 19–20
   for children, 99–100
baleen whales, use of parts in kayaks, 114, 118, 133
Bandi, Hans-Georg, 141–142
beam and beam end, defined, 149
beam seas
   defined, 149
   capsize maneuvers, 21
bearded seals (*Erignathus barbatus*)
   hunting in East Canadian Arctic, 112–114, 115, 127
   use for kayak skin, 35, 112, 134–135, 137
Bech, Kâlêraq, v, 41, 42–43, 45, **51**, 52, 53
Belcher Islands kayak design, 125
beluga whales (*Delphinapterus leucas*), 125–126, 127
Bering Strait (St. Lawrence Island)
   prehistoric kayak designs, 137, **138**, **139**, 140, **140**, 138–142
Bernhardsen, Gaba, 19
bilge stringers, defined, 149
Birket-Smith, Kaj, 13, 62, 63, 67
Birnirk culture (Alaska) kayak designs, 142, 145

bladder floats, use in games, 101
bladdernose or hooded seal (*Cystophora cristata*), 134
blades of paddle, defined, 149, **151**
   See also paddles
Boas, Franz, 125, 127
Borgström, Janne, 75
bow, defined, 149, **150**
bowhead whales (Greenland whales; *arviq/aqviq*; *Balaena mysticetus*), 117–118, 133, 135
bows and arrows
   for hunting caribou, 120, 122
bracing techniques, 21, **55**, 56–57, **57**, 149
Brand, John, 75, 79
breath-holding games, 100
Brielle kayak, West Greenland (1600–1700), 63, **64**, 65
British Arctic Air Route Expedition (1930–1931), 92, **93**, 95–96, 97
British Museum, London, England, 87, 92, 118
Burkhardt, Steve, 34

C

Canadian kayaks. *See* East Canadian Arctic kayaks; Hudson Bay; Labrador
Canadian Museum of Civilization, Gatineau, Quebec, 7, 119, 127, 129
canoes, 112, 124, 127
canted blade forward stroke technique, 49–54
capsize maneuvers, 18
   defined, capsize, 149
   comparison of Alaska and Greenland techniques, 20–21
   early history of, 16–18, 21, 33, 34, 35
   general paddle techniques for, 15
   how to capsize, 22, 25–26
   training routines, 19–20, **20**
   See also capsize maneuvers, standard techniques; capsize maneuvers, storm techniques

capsize maneuvers, standard techniques
Greenland roll, standard (*kinnguffik paarlallugu*),
24, **25**, 24–26
to lie down (*nalaasaarneq*), 29–30, **30**
reverse sweep with paddle across chest (*kingumut naatillugu*), 28–29, **29**
roll with end of paddle in armpit (*paatip kallua tuermillugu illuinnarmik*), 27–28, **28**
roll with paddle held in crook of elbow (*pakassummillugu*), 26–27, **27**
sculling on back (*innaqatsineq*), **21**, **22**, 21–23
sculling on chest (*palluussineq*), **23**, 23–24
capsize maneuvers, storm techniques, 29, **29**, 31–33, **32**
carcasses. *See* hauling and towing carcasses
caribou
lances for hunting, 121–122
hunting of, 50, 102, 118–122
skin and sinew for kayak covers, 134, 135
Caribou Inuit
inland kayak designs, 118–119
kayak designs, 11, 50, 65, 121, 123, 142
paddling strokes, 50
Central West Greenland kayak (c. 1700), 10
Chapelle, Howard, 72, 111
Chapman, F. Spencer, 32–33, 95–96
chest brace, 57, **57**
chest scull, 57
children as kayakers. *See* training
children's kayaks
Southwest Greenland kayak (c. 1613), 11, 83–85, **84**
West Greenland kayak (c. 1889), 67, **68**
chine, defined, 149, **150**
Clachan site, Coronation Gulf, 142
clenched fist roll (*assak peqillugu, qilerlugu/poorlugu*), 39
Clyde River/Kangitugaapik kayaks, 114, **114**, 123
coaming, 76, **76**, **89**, 136
cockpit
defined, 149
cockpit coaming, 76, **76**, 89, 136
evolution and construction (Greenland), 5–8, 10, 11
See also raked cockpit
cockpit rim (hoop), 5–8, 149, **150**
defined, 149, **150**
evolution and construction (Greenland), 5–8
Collins, Henry B., 138
combination roll and brace, capsize maneuver, (*kinnguffik paarlallugu/innaqatsineq*), 31
Comer, George, inland kayak designs, 118–119
construction of kayaks, East Canadian Arctic, 128–137
anthropometric sizing, 131
framework, 129–134, **130**
skin covers, 134–137
tools for, 124–125, 129
wood for, 128–129
See also reconstructions and replicas

construction of kayaks, Greenland
anthropometric sizing, 131
design ancestry, 7–8
early kayaks (17th/18th century), 10
Kâlêraq Bech on Inuit kayak design, 42–43
See also reconstructions and replicas
construction of kayaks, Sweden, 79
Coombs, Howard, 75, 79
Copper Inuit kayak designs, 121, 123, 138, 142
coracle in kayak design ancestry, 137–138
Coronation Gulf kayak designs, 142
courting customs, Greenland, 102
covers for kayaks (*amiq*)
canvas, 35, 112
reconstructions (cloth, fiberglass), 112
skins other than sealskin, 133, 134, 135
See also sealskin kayak covers
Crantz, David, *The History of Greenland* (1767)
capsizing maneuvers described in, 16–18, 21, 33, 34, 35
cross beams, defined, 149, **150**
curved (*avasisaartoq*) kayak, 65–66, **66**, 67

D

Damas, David J., 119
Danish Arctic Expeditions (early 1600s), 11, 15–16
Danish National Museum, Copenhagen, 90, 96, 97, 142
darts, use of, 18, 85, 122, 125–126
Davidsen, Ezekias, 102
Davis, Walter, 42
Davis Strait kayak use, 127
deck, 5, 7, 149
deck beams, defined, 149, **150**
deck strap sculling roll, 35
De Hidde Nijland Stichting, Hindeloopen, the Netherlands, 63
Denmark, recreational kayaking in, 15–16
Disko Bay kayak, 92–93, **93**
disorientation. *See* kayak anxiety and disorientation
Dorset kayak designs, 7, **143**, **144**, 143–145
double kayak designs, 111
Doucette, Vernon, ix

E

East Canadian Arctic, **108**, **109**, **110** (maps)
East Canadian Arctic kayaks, 111–145
Baffin Island kayak, 85, **86**
Clyde Inlet/Kangitugaapik kayaks, 114, **114**, 123
construction of, 128–137, **130**
design ancestry, 137–145
design of, 111–112, **112**, 116
flat-bottomed, 124–128, **125**
influences on Polar Inuit kayak design, 127–128

inland kayaks, 118–122, 125
Labrador kayak, 90, **90**, **91**
North Baffin–North Hudson Bay multichine, 123–124
See also North Baffin Island
East Greenland kayaks. *See* Greenland, history of kayak design
East Hudson Bay kayaks, 125, **125**
Egede, Hans, 16, 41
Ekegrin, Erling, 97
Ekven Cemetery (Chukotka), 138, 141
elbow roll with hand held against neck (*ikusaannarmik pukusuk patillugu*), 40, **41**
entering a kayak, techniques, 20, 87, 100
Eqalulik kayak, Greenland (17th/18th century), 8, 10
exercises, games, and sports, 99–102
exiting a kayak, techniques, 20, 100
expeditions
British Arctic Air Route Expedition (1930–1931), 32, 92, 93, 95–96, 97
Danish Arctic Expeditions (early 1600s), 11, 15–16
James Hall Expedition (1612), 11
Parry expedition (c. 1820), 114, 123
Rasmussen's Second Thule Expedition (1916), 79–80

F

Fabricius, Otto, 16, 18
Ferris, Gail, 19, 41
firearms, impact on kayaking, 66, 126–127, 136–137
fist roll (*assak peqillugu, qilerlugu/poorlugu*), 39
flat-bottomed kayaks
prehistoric designs, 137–138
use in East Canadian Arctic, 124–128, **125**
use in Greenland, 5–7, **6**, 70, 138
floating cockpit rim or hoop, 5–7, 149
floats. *See* bladder floats; sealskin floats (*avataq*)
flutter in paddling strokes, 15, 52–53
Folkens Museum, Stockholm, 79, 80
foredeck, defined, 149
forefoot, 7, 150
foreshaft, defined, 150
Foxe Basin, **108** (map)
inland kayak design, 118–119
kayak design, **112**, 138
Frobisher Bay kayak design, 125

G

Gambell, St. Lawrence Island
prehistoric kayak designs, 138, 139, **140**
games and sports, 99–102
glossary, 149–151
Golden, Harvey, 47, 161

Greenland (Kalaallit Nunaat)
  maps, **viii**, **2**, **3**, **4**
  place names, xi
Greenland, history of kayak design
  Aberdeen, Scotland, kayak (grounded in 1824),
    10, 13–14
  Bech on kayak design, 42–43
  Central West Greenland kayak, 11
  curved (*avasisaartoq*) kayak, 65–66, **66**, 67
  East Greenland kayak, 97, **97**
  East Greenland kayak (c. 1933), 67, 70, **70**
  East Greenland kayak (c. 1960), 70–71, **71**
  East Greenland kayak (c. 1968) (Riel's), 78,
    80–82, **82**
  East Greenland kayak (Chapman's), **95**, 95–96
  East Greenland kayak (Riley's), **91**, 92
  East Greenland white kayak (Watkins's), 93–94, **94**
  evolution and construction, 5–8, **6**, **8**, 131
  Greenland kayaks, 11, 96, **96**
  Polar Greenland kayak (c. 1960), 71–72, **72**
  Skokloster kayak (17th century), 11, 63, 75–78,
    76, **77**, **78**
  Southwest Greenland kayak (c. 1613), 83–85, **84**
  West Greenland child's kayak, 11, 67, **68**
  West Greenland, Eqalulik kayak (17th/18th
    century), 8, 10
  West Greenland kayak (1600–1700) (Brielle), 63,
    64, 65
  West Greenland kayak (1600–1700)
    (Hindeloopen), 63, **64**
  West Greenland kayak (1650–1750) (Hoorn),
    61–63, **62**
  West Greenland kayak (1881), **9**
  West Greenland kayak, 87, **87**, 88–89, **88**, **89**
  West Greenland kayak (c. 1789), 65–67, **66**
  West Greenland kayak (c. 1825), 65, **65**
  West Greenland kayak (c. 1850), 67, **68**
  West Greenland kayak (c. 1916) (Wulff's), 78–80,
    81
  West Greenland kayak (c. 1950), 67, **69**
  West Greenland kayak, 92–93, **93**
  West Greenland kayaks, Danish Expedition (early
    1600s), 11, **12**, 15–16
Greenlanders at Kodiak (video), 57
Greenland Kayak Association (Qaannat Kattuffiat)
  capsizing maneuvers, 18–19, 20–21, 38, 40–41
  championships, 19, 45
  training, 19–20, **20**
Greenland National Museum, Nuuk, 5, 10, 11, 71
Greenland paddles, 14–15, 45–49
  comparison of types of paddles, 45–47, **46**
  construction of paddles, 58
  design ancestry, 63, 72
  feathered *vs.* unfeathered blades, **46**, 46–47
  purposes of design, 14–15, 47
  shouldered *vs.* unshouldered, **48**, 48–49
  sizing of, 14, 47–49, 131

See also paddle design, history of, *and entries
    beginning with* Greenland paddles
Greenland paddles, storm paddles
  defined, **46**, 151
  sizing of, 14, 49
  strokes and uses, 15, 47, 49, 54
  uses of, 54
Greenland paddles, strokes and techniques, 15
  bracing techniques, 55–57, **56**, **57**
  canted blade forward stroke, 49–54, **51**
  cupping the tips, 55
  flutter, 15, 52–53
  paddle extension, 54–55
  side sculling, 56–57, **57**
  stabilizing the kayak with, 47
  training for beginners, 20
  ventilation, 52
Greenland paddles, use in maneuvers
  comparison with Alaska techniques, 20–21
  how to capsize, 22, 25–26
  paddle held lengthwise along spine, hands held
    fore and aft (*aariammillugu*), 35
  paddling upside down with paddle across keelson
    (*pusilluni paarneq*), 35, 37
  reverse sweep with paddle across chest (*kingumut
    naatillugu*), 28–29, **29**
  reverse sweep with paddle behind the neck
    (*kingukkut tunusummillugu*), 35
  roll with end of paddle in armpit (*paatip kallua
    tuermillugu illuinnarmik*), 27–28, **28**
  roll with paddle held in crook of elbow
    (*pakassummillugu*), 26–27, **27**
  sculling roll with paddle across *masik* (*masikkut
    aalatsineq*), 33–34
  sculling roll with paddle held behind back
    throughout roll (*kingup apummaatigut*), 34
  standard capsize maneuver (*kinnguffik paarlallugu*),
    24, **25**, 24–26
  storm roll, capsize maneuvers, **32**, 32–33
  sweep from bow, paddle behind neck (*slukkut
    tunusummillugu*), 35, **36**
  vertical scull (*qiperuussineq/paatit ammorluinnaq*), 34
Greenland roll, standard (*kinnguffik paarlallugu*), **24**,
    25, 24–26, 41
Greenland roll, storm, **32**, 32–33
Greenland whales (bowhead whales; *arviq/aqviq*;
    Balaena mysticetus), 117–118, 133, 135
Gronseth, George, 19, 51, 53
guns. *See* firearms
gunwales
  defined, 149–151, **150**
  evolution and construction (Greenland), 7–8, **8**, 131
Gussow, Zachary, 42

H
hand rolls, Greenland, 38, 39–40, **39**, **40**, 41
Hansen, Ove, 21, **22**, **27**, **29**, **30**, **38**, **39**
harpoons and lances
  defined, foreshaft of harpoon, 150, **150**
  caribou lances, 121–122
  lance with breakaway foreshaft (*anguvigaq*), 113,
    115–118
  as paddle replacements, 37
  use in East Canadian Arctic, 113, 115–118, 122,
    125–127, 137
  use in Greenland, 23, 42, 85, **89**, **152**
  See also caribou; seal hunting; whales and whaling
harpoon line tray or stand
  in East Canadian Arctic kayaks, 137
  in Greenland kayaks, 63, 85
  use in seal hunting, 18, **18**, 23
harp seal (Greenland seal; *qairulik*; *Phoca groenlandica*)
  hunting of, 113
  use for kayak skin and clothing, 100, 112, 134
hauling and towing carcasses
  in East Canadian Arctic, 118, 120, 121, 126, **126**,
    127
  simulation maneuver, 41, 101
Headland, Robert Keith, 61
Heath, John D., v, **v**, 13, 57, 85, 118, 138, 161
Hendrik, Hans, 128
Hindeloopen, West Greenland kayak (1600–1700),
    63, **64**
The History of Greenland (Crantz)
  capsize maneuvers described in, 16–18, 21, 33,
    34, 35
history of kayak design
  prehistoric kayak designs, ix, 5–7, **6**, 137–141, **138**,
    139, **140**, **142**, **143**, **144**
  types of kayaks, ix
  See also East Canadian Arctic kayaks; flat-
    bottomed kayaks; Greenland, history of kayak
    design; paddle design, history of; V-bottom
    kayak designs
hogging, defined, 150
Holm, Pele, 13–14, **14**
hooded or bladdernose seal (*Cystophora cristata*), 134
Hoorn, West Greenland kayak (1650–1750), 61–63,
    62
Hudson Bay, **109** (map)
  kayaks of, 118, 125, **125**, **126**, **133**
  inland kayaks, 119
  North Baffin–North Hudson Bay multichine
    kayak, 123–124, 138
hull, 149
  evolution and construction of (Greenland), 5–8,
    6, 131
Hunterian Museum, Glasgow, Scotland, 11, 65

hunting. *See* bows and arrows; caribou; harpoons and lances; hauling and towing carcasses; narwhal; seal hunting, East Canadian Arctic; seal hunting, Greenland; seals; walrus; whales and whaling
hunting training, 115
hydrofoil, defined, 150

I

Ievoghiyoq site (St. Lawrence Island), 139
Igloolik (Iglulik), 112
Igoolik oral history
   of Aivilingimiut inland kayaks, 118–122
   of hunting, 112–124
   of kayak covers, 135–136
Igloolik Oral History Project, 112, 115
Ikamiut, legend of Parnuna, 41–42
Ikamiut training camp, 19–20, 41–42
ikusaannarmik pukusuk patillugu (elbow roll with hand held against neck), 40, **41**
inboard, defined, 150
inland kayaks, 118–122, 125
innaqatsineq (sculling on back), **21, 22,** 21–23
instruction in kayaking, 19–20, **20,** 99–102, 115
Inuit spelling, pronunciation, and place names, xi
Inujjuaq, sea kayak reconstruction, 112
iqyax (Aleut kayak), 141

J

jackets, kayak (*tuilik*), 20, 30, **30,** 43, 76
James Hall Expedition (1612), 11
Johanson, Lars Emil, 23

K

kajakangst (kayak anxiety), 42
Kalaallit, as place name, xi
Kalaallit (Greenland Inuit) training, games, and sports, 99–102
Kalaallit Nunaat. *See* Greenland
kamiks (boots; *kamiit*), 135–136
Kane, Elisha Kent, kayak collected by, 128
Kangersuatsiaq, West Greenland, kayak, **9**
Kangiqsuk kayaks, 119, **133**
Kangitugaapik/Clyde River kayaks, 114, **114,** 123
Kankaanpää, Jarmo, 145
kayak anxiety and disorientation, 42
   experienced by Ataralaa, 103–105
kayak construction. *See* construction of kayaks
kayak covers. *See* covers for kayaks (*amiq*); sealskin kayak covers
kayak games and sports, 99–102

kayak history. *See* children's kayaks; East Canadian Arctic kayaks; Greenland, history of kayak design; paddle design, history of
kayak jackets (*tuilik*), 20, 30, **30,** 43, 76
kayak maneuvers. *See* capsize maneuvers; Greenland paddles, use in maneuvers; storm techniques, Greenland; rolling maneuvers
kayak paddles. *See* Greenland paddles; paddles
kayak reconstructions. *See* reconstructions and replicas
kayak terminology, xi, 149–151. *See also* paddles, terminology
keelson
   defined, 150, **150**
   in kayak construction (Greenland), 5–8, 7, 131
killer whales, attacks on kayakers, 42
King Island, Alaska, 21, 114
kingukkut tunusummillugu (reverse sweep with paddle behind the neck), 35
kingumut naatillugu (reverse sweep with paddle across chest), 28–29, **29**
kingumut naatillugu/palluussineq (Greenland storm roll, capsize maneuvers), **32,** 32–33
kingumut naatillugu/palluussineq (storm techniques, recovery combination, capsize maneuver), 29, 31–32
kingup apummaatigut (sculling roll with paddle held behind back throughout roll), 34
kinnguffik paarlallugu (Greenland roll, standard capsize maneuver), **24,** 24–26, **25, 41**
Kiyuaqjuk (kayak maker), 119
Kleist-Thomassen, Hans, 18, 67
Koryak (Siberia) and East Greenland kayaks, comparison of, 5, **6,** 7
Kotzebue Sound, Alaska, capsize maneuvers, 21
Kylsberg, Bengt, 75

L

Labrador, Canada, **110** (map)
   kayaks of, 90, **90, 91,** 126
   wood for kayak making, 129
Lagerkrantz, Christian, 79
lakes, use of kayak in, 118–122, 125
lances. *See* harpoons and lances
Langgård, Per, 18
legends, 41–42
Lindsay, Martin, 95
line sports (Greenland Inuit), 101–102
longitudinals, defined, 150
loom, defined, 150, **151**
Loring, Stephen, 138
Lübeck, Germany, kayak, Danish Arctic Expeditions (early 1600s), 11, **12,** 77
lying down in kayaks
   as capsize maneuver (*nalaasaarneq*), 14, 29–30, **30**
   resting or sleeping in kayaks, 22, 114, 124

M

MacDonald, John, 112
Mackenzie Inuit kayak designs, 142
making kayaks. *See* construction of kayaks
maneuvers, kayak. *See* capsize maneuvers; Greenland paddles, use in maneuvers; rolling maneuvers; storm techniques, Greenland
maps, **viii,** 2, **3, 4, 108, 109, 110**
Marischal Museum, Aberdeen, Scotland, 10, 11, 13
Mary-Rousselière, Guy, 117, 143–145
masik, defined, 150, **150**
masikkut aalatsineq (sculling roll with paddle across *masik*), 33–34
Mathaeussen, Kristoffer, 26
Mathaeussen, Manasse, v, 10, 14, 19, 22, 26, 30, 40, 41
Mathiassen, Therkel, 128
McClintok, Francis Leopold, 128
McGregor, John, 79
Medico-Chirurgical Society, Aberdeen, Scotland, kayak, 11
men's roles in kayak construction, 135–136
Mikkelsen, Ejnar, 13
Mittimatalik (Pond Inlet), xi, 112
   Dorset kayak designs, **143,** 143–145, **144**
   reconstruction of kayak (1973), 117
Miyowagh site (St. Lawrence Island)
   and prehistoric kayak designs, 138, **138**
Moore, J. I., 32
Mount Ingik, Greenland, 41–42
multichine hulls, East Canadian Arctic, 118, 123–124, 138
Museon, the Hague, the Netherlands, 70

N

nalaasaarneq (lie down as capsize maneuver), 29–30, **30**
narwhal (*qilalugaq; Monodon monoceros*)
   hunting in East Canadian Arctic, 112–114, 115–118, 125–126, 127, 128
Nasook, Martha, 135–136
National Museum of the American Indian, New York, 7
nerfallaallugu assakaaneq 10 sekunit (Greenland roll), 41
Netsilik Inuit kayak designs, 121, 123, 138, 142
Newfoundland Museum, 125
Nicodemossee (kayak maker), 92, 93, 95
noise
   kayak design to reduce, 124
   sealskin covers to reduce, 135–136
   use in walrus hunting, 116
Nooter, Gert, 10, 61, 63, 70
Nootka canoe, 124
norsamik kingukkut (throwing stick, leaning aft), 38

norsamik masikkut (throwing stick swept outward and down from near *masik*), 38, **39**
norsamik nerfallaallugu (throwing stick swept aft, then forward), 38–39
North American Arctic, **viii** (map)
North Baffin Island, 123
  Dorset kayak designs, **143**, **144**, 143–145
  hunting caribou on inland water, 118–121
  hunting sea mammals on, 112–118
  Kangitugaapik/Clyde Inlet kayaks, 114, **114**, 123
  North Baffin–North Hudson Bay multichine kayak, 123–124, 138
  oral history project, 112, 115
  See also Baffin Island; East Canadian Arctic kayaks; Igloolik oral history
North Hudson Bay
  North Baffin–North Hudson Bay multichine kayak, 123–124, 138
Nungak, Zebedee, 112
Nungavik site (Baffin Island), 143–144
nusutsinneq (simulation of seal dragging), 41
Nuuk Kayak Club Collection, 67

O

Okvik–Old Bering Sea kayak designs, 137, 138, 139, 141
Olsen, Simon, 42
Olsen, Steffen, 49
outboard, defined, 150
outrigger, defined, 151

P

paddles, 37, 45–49
  blades, 45–47, **46**, 49, 50, 149
  comparison of types of, **46**, 46–47
  feathered *vs.* unfeathered blades, **46**, 46–47
  sizing of, 47–49, 131
  use of, 47, 117, 118, 121
  See also Greenland paddles; paddle design, history of; paddles, terminology
paddle design, history of, 5
  prehistoric paddles, 137, **142**, 142–143
  in East Canadian Arctic, 85, 112, 117, 124
  in East Greenland, 91, **91**
  in Southwest Greenland, **84**, 85
  in Swedish kayak tradition, 77, 80, 81, **81**, **82**
  in West Greenland, 87, **87**, 88, **89**
  See also history of kayak design
paddles, terminology, 149, 150, 151, **151**
  flutter, 52–53
  sculling, 56, 151
  ventilation, 52
Padilla, Maligiaq, v, 45, 48, 49, 50, 52, 53, 54, 57

pakassummillugu (roll with paddle held in crook of elbow), 26–27, **27**
pallortillugu assakaaneq 10 sekuntit (roll forward, hands in normal position), 40
palluussineq (sculling on chest), 21, **23**, 23–24
palluussineq variation (sculling on the chest used as a roll), 34
Parnuna, legend of, 41–42
Parry, William E.
  kayak collected by, 114, **114**, 123
  kayak observations of, 118
pautik, defined, 151
Petersen, H. C., 8, 10, 48, 49, 62, 65, 134
Petersen, John, 34, 40, 41, 49, 57
Peyrère, Isaac de la, 61
Piugaattuk, Noah, 115–118, 119–122, 124
Pivat, Ulrik, 21
place names, Inuit terms, xi
polar bears, 115
Polar Inuit (Greenland), introduction of watercraft, 71–72, **72**, 127–128
Pond Inlet. *See* Mittimatalik (Pond Inlet)
port, defined, 151
portage for kayaks, 102, 119, 124
Poulsen, Thimothaeus, 10
prehistoric kayak designs, 5–7, 137–145, **138**, **139**, **140**, **142**, **143**, **144**
prehistoric paddle designs, 5–7, 137, **142**, 142–143
pulling carcasses. *See* hauling and towing carcasses
Punuk culture (St. Lawrence Island) kayak designs, **138**, **139**, **140**, 138–142, 145
pusilluni paarneq (paddling upside down with paddle across keelson), 35, 37
Puvirnituuq-Inujjuaq kayaks, 125

Q

qairulik (harp seal; *Phoca groenlandica*)
  hunting in East Canadian Arctic, 113
  use for kayak skin and clothing, 100, 112, 134
Qajaq Club, Nuuk, 10, 19–20
Qajaq USA, 58
qasuersaartoq (the one resting), 29
qayaq/kayak terminology, xi
Qeqertasussuk (Saqqaq site) kayak designs, 137
qilalugaq (narwhal; *Monodon monoceros*)
  hunting in North Baffin, 112–114, 115–118, 125–126, 127, 128
qiperuussineq/paatit ammorluinnaq (vertical sculling roll), 34
Quist, Ludwig, 15

R

Rae, John, kayak observations of, 118
raked cockpit, 7–8, 151

Ranshaw, Charles, 11, 85
Rasmussen, Knud
  gift of East Greenland kayak, 97, **97**
  records Ataralaa's narrative (1899–1900), 103
  Second Thule Expedition (1916) of, 79–80
reconstructions and replicas
  Aivilingmiut kayaks (1967 and 1968), 119
  Akudnirmiut kayak (1970s), 127
  East Canadian kayaks (1980s), 112
  Nungak, great sea kayak (2000), 112
  Pond Inlet, Mittimatalik kayak (1973), 117
  purposes of, 77, 95–96, 112
  West Greenland kayaks, 88, 92–93, **93**
reverse sheer, defined, 151
reverse sweep
  with paddle across chest (*kingumut naatillugu*), 28–29, **29**
  with paddle behind the neck (*kingukkut tunusummillugu*), 35
ribs, defined, 150, **151**
Richter, Søren, 142
Riel, Jörn,
  East Greenland kayak of, 78, 80–82, **82**
Rijksmuseum voor Volkenkunde, Leiden, the Netherlands, 10, 63, 67
Riley, Quintin, 87
  kayak of, **91**, 92
ringed seals (*Phoca hispida*), 112, 126, 134
Robertson, Thomas, 67
rocks, use with rolls
  clenched fist roll (*assak peqillugu, qilerlugu/poorlugu*), 39
  rolling with a rock swept outward and down (*ujaqqamik tigumisserluni*), 39–40
rolling maneuvers, East Canadian Arctic, 116
rolling maneuvers, Greenland
  Greenland storm roll, **32**, 32–33
  rolling with arms crossed (*tallit paarlatsillugit paateqarluni/masikkut*), 35, **36**
  roll with end of paddle in armpit (*paatip kallua tuermillugu illuinnarmik*), 27–28, **28**
  roll with paddle held in crook of elbow (*pakassummillugu*), 26–27, **27**
  standard Greenland roll (*kinnguffik paarlallugu*), 24, **25**, 24–26
root of blade, defined, 151, **151**
rope gymnastics, 19–20, **20**
Rosing, David, 48
Rosing, Johannes, 31, 103–105
Ross, John, 114
Royal Albert Memorial Museum, Exeter, 114
Royal Ontario Museum, Toronto, Ontario, 7, 125, 133
Royal Scottish Museum, Edinburgh, 133
Rymill, John, 32

S

sagging, defined, 151
Saladin, Bernard, 119
Sallirmiut kayak designs, 123
Samuelsen, Karl, 19, 23, 42
Sandbord, Harald, 33
Saqqaq kayak designs, 137
Schiffergesellschaft, Lübeck, Germany, 11
Schmidt, Jan, 15
sculling and sculling rolls, Greenland
    defined, 56, 151
    deck strap sculling roll, 35
    sculling on back (*innaqatsineq*), **21**, **22**, 21–23
    sculling on chest (*palluussineq*), **23**, 23–24
    sculling on the chest used as a roll, 34
    sculling roll with paddle across *masik* (*masikkut aalatsineq*), 33–34
    sculling roll with paddle held behind back throughout roll (*kingup apummaatigut*), 34
    side sculling, 56–57, **57**
    vertical scull (*qiperuussineq/paatit ammorluinnaq*), 34
seal hunting, East Canadian Arctic
    flat-bottom kayak design and, 126
    hauling and towing carcasses, 120, 121, 126, **126**, 127
    in North Baffin Island, 112–114, 119–121
seal hunting, Greenland, 15, 18, 41–42
    Ataralaa's kayak trip (1899–1900), 103–105
    and kayak design, 42–43
    capsize manuevers for, 23
    See also harpoons and lances; hauling and towing carcasses
seals, 112, 134. *See also* bearded seals, harp seals, seal hunting
sealskin floats (*avataq*), 139
    construction and uses, 37, 117
    use in games and exercises, 101
    use in Greenland, 9, 16, 23
    use in hunting, 113–117, 120–121
    use in hunting in East Canadian Arctic, **118**, 126, 139
    use in roll (*avataq* roll), 37–38, **37**, **38**
sealskin kayak covers
    in East Canadian Arctic, 112, **134**, 134–137
    in Greenland, 7–8, 35
Second Thule Expedition, Rasmussen (1916), 79–80
Seklowaghyaget site (St. Lawrence Island)
    and kayak designs, 139, 140, **140**, 141
sheer or sheer line, defined, **150**, 151
Siberian and East Greenland kayaks, comparison of, 5, **6**, 7
Sicco harpoon heads, 142
side sculling, 56–57, **57**
Silis, Ivars, *Amphibious Man* (video), 49
Singhertek, Henrik, 70
Sisimiut Museum, 10

siukkut pallortillugu (hands in normal position, roll forward), 37
sizing (anthropometric)
    of Greenland paddles, 14, 47–48, 49, 131
    of kayaks, 131
skeg, defined, 151
skin sewing, **134**, 135–136
    See also sealskin kayak covers
Skokloster kayak (17th century), 11, 63, 75–78, **76**, **77**, **78**
sleeping in kayaks, 22, 114, 124
slukkut tunusummillugu (sweep from bow, paddle behind neck), 35, **36**
Smith, Carl, 79
Smithsonian Institution, 7, 138
Souter, William Clark, 11, 61
South Hudson Strait kayaks, **133**
spatial disorientation, 42
speed, 102, 121, 123–124, 126
sports and games, 99–102
sports kayaks
    comparison of paddle designs, 45–47, **46**, 49
    plastic imports, 112
    recreation in Denmark, 16
    See also reconstructions and replicas
St. Lawrence Island, 138
    prehistoric kayak designs on, **138**, **139**, **140**, 138–141
    and whaling and Punuk warfare, 141–142
stall, defined, 151
stands for kayak storage. *See* storage of kayaks
starboard, defined, 151
Steensby, H. P., 62, 72
stem, defined, **150**, 151
stern, defined, **150**, 151
storage of kayaks, 9, **125**, **130**, 137
storm paddles. *See* Greenland paddles
storm techniques, 29, 31–33, **32**
straightjacket roll (*tallit paarlatsillugit timaannarmik*), 40
stringers, 5–8
    defined, 150, **150**, 151
surf-riding exercise, 100
Surgeons' Hall, Edinburgh, Scotland, kayak, 11
Sutherland, Patricia D., 145
svimmelbedden (dizziness), 42
swede form (design term), 79
Sweden, 75–82
    kayak tradition of, 78–79
    Riel's East Greenland kayak, 78, 80–82, **82**
    Skokloster kayak (17th century), 11, 63, 75–78, **76**, **77**, **78**
    Wulff's West Greenland kayak, 78–80, **81**
sweep from bow, paddle behind neck (*slukkut tunusummillugu*), 35, **36**
swells, defined, 151
syllabic system, Inuit (Canadian), xi

T

Tagurnaq (kayak maker), 119
tallit paarlatsillugit paateqarluni/masikkut (rolling with arms crossed), 35, **36**
tallit paarlatsillugit timaannarmik (straightjacket roll), 40
terminology, kayak, xi, 149–151
    See also paddles, terminology
Thomsen, Thomas, 71
Thorell, Sven, 79
throwing stick (*norsaq*), 18, 38, **39**, 85, **89**
throwing stick rolls (Greenland), 38–39, **39**
Thule culture, 142, 143
    kayak designs of, 7, 20–21, **142**, 142–145
thwarts, defined, **150**, 151
Tinbergen, N., 67
Tittussen, Luutivik, **35**
Tobiassen, Pavia, 21, **40**
tools for making kayaks, 124–125, 129
towing carcasses. *See* hauling and towing carcasses
Town Hall, Braintree, Essex, England, 92
trade, and kayak designs, 127, 128, 129, 136
training, 19–20, **20**
    of children, 99–102, 115
Trinity House, Hull, England, 11, 83
tuilik (waterproof kayak jacket), 20, 30, **30**, 43, 76
Tununirmiut (Aggumiut) multichine kayak design, 124
Turner, Lucien M., 125, 128
Tyrrell, James W., 133

U

ujaqqamik tigumisserluni (rolling with a rock swept outward and down), 39–40
ujjuk (bearded seal; *Erignathus barbatus*)
    hunting in East Canadian Arctic, 112–114, 115, 127
    use for kayak skin, 35, 112, 134–135, 137
umiak (pl. *umiat*), ii, 102, 117–118, 128, 136
    prehistoric, 137–141, **143**, **144**, 143–145
    walrus skin covers for, 134
Ungava Bay, **109**, **110** (maps)
    kayaks, 119, 125, 128, 134
University of Cambridge Museum of Archaeology and Anthropology, 85, 88
Uppsala, Sweden, Skokloster Palace, 11
Uyarasuk, Titus
    on hunting in North Baffin, 112–114, 115, 123–124
    on women's sewing of skins, 136

V

varnish for sealskin covers, 136, 137
V-bottom kayak designs
    East Canadian Arctic kayaks, 124
    Greenland kayak tradition, 65, 78, 123
    Swedish kayak tradition, *77–78*, 80
ventilation in paddling technique, 52–53
vertical scull (*qiperuussineq/paatit ammorluinnaq*), 34
vortex shedding, 15

W

walrus (*aiviq*)
    hunting of, 115–120, 124, 127, 141
    use of skin for kayak covers, 134
walrus-pull technique, 58
warfare, Punuk, 141–142
Watkins, Gino, white kayak of, 34, 93–94, **94**
Watkins expeditions (1930s), 32
weapons and kayak design, 125
West Canadian Arctic–North Alaska kayak designs,
    123
Westfries Museum, Hoorn, the Netherlands, 61
West Greenland, **3** (map)
West Greenland kayaks. *See* Greenland, history of
    kayak design
West Hudson Bay kayak designs, **118**
whales, baleen, use in kayak construction, 114, 118,
    133
whales, beluga (*Delphinapterus leucas*), **118**, 125–126,
    127, 135
whales, Greenland or bowhead (*arviq/aqviq*; *Balaena
    mysticetus*), 117–118, 133, 135
whales, killer, attacks on kayakers, 42
whales, white. *See* whales, beluga
whales and whaling, 72
    in East Canadian Arctic, 117–119, **118**, 121,
        125–127, 133
    influence of trade on kayak designs, 129
    and Punuk culture, 141–142
    use of parts in paddles and kayaks, 72, 114, 118,
        133, 134
    in West Canadian Arctic, 141–142
Whitaker, Ian, 13
Whitby Museum, England, 65
Wilkes, R. Jeffrey, 50
Wilkinson, Douglas, 126
women's roles in kayak construction, **134**, 134–136
wood for construction of kayaks and *umiat*
    in East Canadian Arctic kayak and *umiak* designs,
        117, 128–129
    in Greenland kayak and paddle designs, 7, 14, 58
Wrangel Armoury, Skokloster Palace, Sweden,
    kayak (17th century), 11, 75
Wulff, Thorild, West Greenland kayak of, 78–80, **81**

Bent Frederiksen and Jørgen Aronsen of Nuussuaq, Upernavik, August, 1968.
Courtesy Keld Hansen.

About the Authors

E. Arima is a Canadian ethnologist working in the Arctic and Northwest Coast culture areas. Since 1975 he has been an ethnohistorian with Parks Canada. In addition to his studies of watercraft, he has worked on Inuit oral traditions of East and West Hudson Bay, Kwakiutl mask carving, and Blackfoot history.

John Brand is a world-renowned expert on the history of kayaks; he has surveyed historic kayaks housed in museums in Denmark and throughout England. He is the author of *The Little Kayak Book* series. He lives in Colchester, Essex.

Hugh Collings has always been interested in long, narrow craft, whether single sculls, kayaks, or sailing canoes. He has built several replicas of Greenlandic kayaks and has paddled across the Baltic Sea from Turku, Finland to Stockholm, Sweden. He lives with his Swedish wife in the small coastal town of Oxelösund south of Stockholm where he has ample opportunity for boating in the magnificent archipelago.

Harvey Golden has been conducting research on the forms and construction of arctic kayaks for five years. He has traveled to numerous museums in the United States, Canada, Greenland, and Europe. He has also built and used over forty full-size replicas of different kayak types in order to gain greater insight into their design.

John Heath (1923–2003) was one of the world's foremost authorities on traditional arctic watercraft. He was dedicated to preserving and promoting knowledge of kayak structure, design, and technique among both scholars and recreational kayakers. He conducted research on kayaks at museums throughout Europe, Greenland, and North America and published numerous articles in *American White Water* and *Sea Kayaker*.

Greg Stamer is a long-time student of Greenland kayaking techniques and credits John Heath and Greenland champion Maligiaq Padilla as his primary mentors. He has competed twice in the Greenland National Kayaking Championship, winning his age class overall in 2000 and the rolling event for his age class in 2002. He is the president of Qajaq USA (www.qajaqusa.org), the American chapter of the Greenland Kayaking Association.